The Death of Bruce Lee:

A Clinical Investigation

The Death of Bruce Lee:

A Clinical Investigation

By

Duncan Alexander McKenzie R.N.

The Death of Bruce Lee:
A Clinical Investigation
Published by the author.

ISBN 978-1-300-10886-3

Book formatted by www.bookformatting.co.uk.

Contents

Foreword

His name was synonymous with, strength, speed, athleticism and vigor, and he had been called the fittest man in the world, and pound for pound, the strongest man on Earth. The physical feats that he could perform were legendary. It was reported that his speed, in terms of reacting and punching from a distance of three feet away, was determined to be around five hundredths of a second (0.05 second); from five feet away it was around eight hundredths of a second (0.08 second).[1] He could take in one arm a 75 pound barbell and in a standing position, with the barbell held flush against his chest, slowly stick his arm out and then lock it, holding the barbell there for several seconds. In order to demonstrate the speed at which he could move, Lee could snatch a dime off a person's open palm before they could close it, and leave a penny behind. He performed one-hand push-ups using only the thumb and index finger and could complete 50 reps of one-arm chin-ups. In order to develop and maintain his abdominal muscles, which he felt were of critical importance in the martial arts, he would hold an elevated v-sit position for 30 minutes or longer. That is, from a seated position raise the legs to the point where they are 45degrees or more above the horizontal, and hold them there for that period of time. His punches and kicks were reported to be explosive. He could cause a 300-lb (136.08 kg) bag to fly towards and thump the ceiling with a sidekick. Such was the strength, power and fitness of the man who was described as a 'kinetic genius.'[2] The martial artist Ed Parker described him, in terms of his abilities, as a 'one in two billion' person.[3] Others praised him also. 'Bruce Lee was blessed with good long arms, incredible initiation speed and reaction time, and he hit big-hard compared to others in the Asian fighting disciplines. Bruce had phenomenal attributes,' confirms Joe Lewis. 'Amazing speed, power, strength, reflexes.' Kareem Abdul Jabbar, the American basketball player, was more than a little impressed with Lee's prowess. 'He had cunning, killer instinct and the will to dominate. And incredible athletic skills: balance, eye-hand coordination, and timing. And all that driven by a very intense will.'[4] Yet at the age of 32 years, this extraordinarily fit and strong man suddenly died. Of reportedly natural causes.

Bruce Lee stormed into the consciousness of the world through the movies that he made and he possessed a charisma and an intensity that was unparalleled. When he died I was completing my first year of nurse training in the United Kingdom. As everywhere else his movies had been phenomenal blockbusters in that country, and as the pop song so succinctly stated, 'everybody was Kung Fu fighting.' In the country where I now live, the Philippines, he was equally successful and his movie 'Fist of Fury' was so popular that the government had to place restrictions on such foreign imports. This was in order to protect their domestic film producers.[5] At a press conference, Manny Pacquiao, the Filipino boxing champion, said that "Bruce Lee is my idol," and he's not the only person in the world to share the sentiment. Lee has legions of fans spread across the globe, and as the years pass, the collection only grows.'[6] At the time of his death and beyond people were drawn to martial arts schools in droves, myself included. In the United Kingdom I began karate training and I later took this up again at the age of 35 after migrating to Australia. I studied under Shihan Kurt Klimkait in Shukokai Karate in Melbourne and obtained my black belt after 5 years of hard training alongside others such as Richard Callaghan, who is now a highly graded karateka and a talented athlete and martial artist. Believe me, under Shihan Klimkait it was hard, no-nonsense training. This style, by the way, was founded by Shihan Tani and one of his principal students was the late Shihan Kimura, whom I trained under when he travelled from the States to give training courses in Australia. I also attended training courses in Melbourne given by Shihan Chris Thompson, a South African karateka, and Shihan Tommy Morris who hails from my country of birth, Scotland. Shukokai karate prides itself on the development of hard impact of punches and kicks. Shihan Kimura, like Shihan Klimkait, could kick like a mule. Oh, the happy memories of flying through the air while holding the impact pad for Shihan Kurt. For a small guy like myself there was one golden rule in sparring when faced with Shihan Klimkait. And that was, *move out of the way fast!* But it was fun and it was deadly serious and it was an important part of my life. Unfortunately I developed rheumatoid arthritis, which put an end to my martial arts training. This was also one of the factors that led to me retiring early from nursing and moving to the tropical climate of the Philippines, from the bone-chilling winters of Melbourne. However the wonderful memories of the training remain. I was very much an average karateka. Someone like Bruce Lee would have had more raw talent in his little toe, than I had in my whole body. But I worked hard. Bruce Lee and his extraordinary skills, talent and charisma were things that attracted us all to the martial arts. I mention this because, by virtue of my karate training, I have some insight into the martial arts, and the capacity to sort the truth from fiction. The martial arts, karate or kung fu or judo or whatever, do not confer invincibility on its practitioners. In combat, of any description, there are only probabilities that someone will win based on factors such as strength, size, weight, technique, age, and the all-important psychological attributes of determination and the capacity to absorb and deal with pain, and fear. And yet we still see in the literature today endless, and meaningless, debate about whether Bruce Lee could be beaten in a fight. For anyone who has been involved in the martial arts the answer would be simply, 'Of course he could be beaten!' And that is because as Bruce Lee aged his reflexes would have slowed, and the advancing years would have robbed him, as it does all of us, of his vigor and his strength. He could be beaten if he was faced with a number of opponents and he could be beaten if he had lived to an advanced age and someone much younger had sparred with him. And perhaps, even in his prime, he could have been beaten by Joe Lewis or someone else. We don't know, and it doesn't really matter. Because that, actually, is not what the martial arts are about. That extraordinarily talented martial artist Chuck Norris has been asked on many an occasion if he had sparred with Bruce Lee and if he had beaten him. Chuck Norris is far too much of a gentleman to have ever gotten into that kind of debate, and had too much respect for Bruce Lee, and would always deftly sidestep the issue by stating that they just worked out together. Bruce Lee was human. He was not invincible. He was not endowed with special, superhuman powers. But his intensity and his charisma and his skill and his prowess were really quite something else.

Bruce Lee rose from being an ordinary guy to being a superstar, but the dream of his stardom for him abruptly ended, all too soon after it had begun. However, we are left with his enduring legacy and a timeless testament to the perfection of the human body and the extraordinary physical skills that can be developed and honed when one has the drive and the raw talent. This book is not about the life of Bruce Lee. Of that we have so many. It is about his medical history and his death. It is about how exactly he died, and we go here deeper into the information that we have been given to fully analyze what went wrong and I ask the question, was his death avoidable? For this we have to go deeply sometimes into medical issues, but that is necessary. And so this is, by necessity, a clinical treatise, and is not light reading. However it has to be this way for we have to review the precise science of the issues at hand and what the exact mechanisms are in regards to the medical issues that we are discussing.

As a nurse I have an abiding interest in all things clinical and in forensic medicine. And so here, in this book, is the truth as *I* see it, about his life and untimely death simply as a Bruce Lee fan and as a registered nurse. Therefore, as someone with some clinical knowledge and expertise. It is from the perspective of an old registered nurse who has dealt with the sorts of things that Bruce Lee died of. This book is fully researched and fully referenced and everything that I use to build my arguments and conclusions on are fully sourced. When we write a biography, and wish it to be an *exact* biography, I personally consider it necessary to reference the work properly. All of what I write will be open to public and professional scrutiny. By Bruce Lee fans and by medical professionals and anyone else who may be interested. This is how knowledge and truth advances. I see no place in this debate for hysteria, unreasoned argument or personal insult. It is about the biography of a man called Bruce Lee and his death, and what the truth actually is concerning that. Everyone who writes a book and makes statements about a matter can and should be challenged. This is completely good and normal and constructive.

We owe it to the memory of the dead to protect them from that which may harm their memory or legacy. As Bruce Lee so famously stated in one of his films when a *karateka* broke a board in a display of strength in front of him, "Boards don't hit back." And neither can the dead hit back. And so we owe it to Bruce Lee, who gave us so much, to set the record straight if this is required. I am a great admirer and fan of Bruce Lee and thus I must state my bias at the beginning. However, I will be as objective as possible. This book, like any other, is a journey. By writing it I have learnt a lot, and I hope you will learn from it too. And if in some way it brings you close or helps you understand better one of the greatest martial artist of all time, and his untimely death, then I will have achieved my aim. This book goes into medical matters quite deeply. As I stated above this is necessary in order to comprehensively deal with the subject matter and to fully explore the issues of his death. It will therefore not appeal to everyone, as it is not simply the story of his life and death but an attempt to explain what happened from a clinical point of view. For health professionals it may therefore prove of interest in relation to the forensic medical matters discussed, and such issues as adverse medication reactions. For those Bruce Lee fans who want to understand precisely why he died, as far as we can determine, then this book I hope will take its place amongst the collection of Bruce Lee biographies already published. This one having a specific clinical and medical focus.

This book as I have stated is fully referenced and this is the way I do my research and writing. Referencing should be a part of any research document or biographical work. Good research involves ensuring that, 'Any factual material or ideas you take from another source must be acknowledged in a reference, unless it is common knowledge.' And further, 'Your method of referencing must tell your reader where you got all the specific information in your paper, and where any ideas or interpretations came from that are not your own thinking.'[7]

During my research for this book I came across, in digital form, a photograph that is purportedly of Bruce Lee in the morgue after his death. I emailed the source of this photo to enquire as to the authenticity of this picture. He was unable to vouch categorically for its authenticity. The next step, had someone been able to vouch for its authenticity, would have been to check as to determine if there were any copyright issues in terms of its use. In view of the fact that I was not able to even get past the authenticity stage, I have not included it in this book. I find the picture sad and disturbing to look at. To me, it *appears* to be him. This once dynamic, extraordinary man now lifeless on a mortuary slab. However, this book is not about sensationalism or voyeurism. It is about the truth and I have tried to verify and source everything so that we can arrive at reasonable and logical conclusions, and to treat the matter in a respectful and professional way. This is the way I write and research.

And so was he poisoned? Was he a victim of the *dim mak* death touch? Did drugs kill him? And what were the circumstances of his final days and moments, particularly from a medical point of view? Did he have an undescended testicle? Was he on steroids? Did he have a serious back injury? Was he a drug addict? Why was the man who was referred to as the fittest and strongest man in the world rejected on health grounds from the US military during the Vietnam draft years? What really killed him? In this book we will examine these issues and answer these questions, as best we can, in a methodical and rational manner, and we will go deeper into the medical issues. For anyone with an interest in medical matters and forensics, which is defined as relating to or dealing with the application of scientific knowledge to legal problems, then this book is just for you!

FOREWORD

Before we begin on this journey towards an attempt at an understanding of what really happened, there is another thing that must be said. That is, in relation to the death of Bruce Lee there are only a few incontrovertible facts. One is that he is actually dead. Believe or not some have clung to the belief that his death was faked. The second is that the *immediate* cause of death was cerebral edema. That is an incontrovertible fact and was, to repeat, the *immediate* cause of his death. However, we must logically ask, what actually caused the cerebral edema? The third incontrovertible fact is that at this point in time there is no specific *unequivocal* way to determine what exactly the cause of the cerebral edema was. Regarding this we must work on the basis of a hierarchy of possibilities, or probabilities. And this I will do, sifting each one to determine what the possibility is of it actually providing an explanation of the cerebral edema that caused his death. In the end we will be left with the most probable cause, and whether it is a sensation or whether the truth is simply mundane, that is where we will go because we are seeking the truth, as best as we can determine it to be.

As a preface to this book it is important to put something else clearly in context. Cultural perspective is always a critical issue in the biography of any man or woman. We are born into a specific cultural matrix and view the world and 'reality' through that. And this cultural matrix shapes our behavior. Bruce Lee, although born in San Francisco in the United States, spent all of his formative years in Hong Kong. He grew up with a completely Oriental cultural matrix. It wasn't until he was 18 years of age that he returned to the States. Within this Oriental matrix the use of substances such as Opium for instance were viewed very differently than in the West. Reportedly, Lee's father used the opium pipe as did Lee's martial arts instructor Yip Man.[8] It was a commonplace and openly accepted practice. Concubinage was another culturally accepted phenomenon. 'Concubines are women who cohabit with men but are not married to them. In ancient China it was common for successful men to have several concubines – the Chinese Emperors often kept thousands. Concubines' situation ranged from pseudo-wives to poorly treated prostitutes. Concubines do not officially exist in modern China, but 'Ernai' or 'second wives' are increasingly common. Unlike in the West, keeping a mistress is not always frowned upon in China. The CCP [Chinese Communist Party] tried to stamp out concubinage, which they saw as a feudal vice, *but among China's new breed of super-rich businessman, keeping a young, fashionable, spoilt young woman as a mistress can gain you face – which in turn is good for business.* Concubinage was not abolished in Hong Kong until 1971.'[9] These cultural norms and traditions must be taken into account when we look at the life of Lee, and more specifically his death and the circumstances of his death. And he should not be judged or censured from the view of a Western Judeo-Christian cultural standpoint.

So let's get underway and see if there is anything new, or different, that we can add to the historical record and the biography of this astonishing man called Bruce Lee. History demands that individuals be judged according to the facts of their lives, without prejudice as far as that is possible. Enjoy the journey!

Acknowledgements

I was motivated to write this book after reading 'Unsettled Matters: The Life and Death of Bruce Lee' by Tom Bleecker. Mr. Bleecker and I hold differences of opinion in regards to certain matters regarding Bruce Lee and his life and medical history, and Bruce Lee's death. However Mr. Bleecker showed great courtesy and respect towards me by kindly communicating with me over certain matters of contention, and offering constructive comment and criticism of this book prior to its publication. I am sincerely grateful to him for this, and for his book which details many issues of great significance that are essential for an understanding of this extraordinary man called Bruce Lee.

THE DEATH OF BRUCE LEE: A CLINICAL INVESTIGATION

Chapter 1
Theories of how Bruce Lee died.

"If I should die tomorrow, I will have no regrets. I did what I wanted to do. You can't expect more from life."Bruce Lee.[1]

Since the death of Bruce Lee on July 20, 1973 rumor, speculation, theory and conjecture have been the constant companions of his life story. Such was the mythos of the man and his extraordinary talent that it would seem somehow unfitting that he would die of a simple cause, one that was somehow without mystery or sensation. Later in this book we will outline how he died, using the facts that we have and the evidence that is available. The theories and speculation that surround his death relate to the following claims:

Bruce Lee was allegedly murdered.

Allegedly the main suspects tied up in this speculation is that he was murdered by the triads in Hong Kong, or that he was murdered by a group of oriental martial artists, or a single martial artist, who disliked the fact that Lee was giving away the 'secrets' of Kung Fu (Gung Fu) to non-Orientals. The method of his murder supposedly revolved around two methods. One was poisoning and the other was by the use of the mythical *dim mak* touch that certain Kung Fu practitioners allegedly possess. The first assertion, that he was killed by the Triads in Hong Kong, is easily disposed of. The autopsy of Bruce Lee, which we will detail later, revealed no physical injuries of note. The Triads are usually ruthless and they would kill anyone who crossed them, and usually in a not-too-subtle manner, to serve as a warning to others. 'In addition to grotesque beatings, actors were kidnapped and actresses raped for refusing to cooperate with triads. Knife slashings were commonplace, limbs were severed, agents were assassinated, and fire bombings were not infrequent.' So we are advised in 'Unsettled Matters'. During the time just before Lee's death some have commented that Lee suffered from depression and 'paranoia'. In fact when someone is in such a dangerous and threatening environment they cannot be called 'paranoid'. They actually have real enemies! And so I view Bruce Lee as being under significant stress, but he was not paranoid. You had to be on your toes in such an environment and hypervigilant. Some have maintained the view, that he was paranoid, but then at the same time they hold the view that he may have been poisoned! Therefore just because an individual is paranoid does not necessarily mean his or her suspicions are false, as noted in the movie Catch-22: "Just because you're paranoid doesn't mean they aren't after you."[2] In actual fact, as the financial affairs reveal after the death of Bruce Lee, he was actually far too *trusting* of others. He had left far too much in the hands of others, trusting them when he should actually have been protecting his own financial interests. At the very end of his life he was given a tablet for a headache by Betty Ting Pei. We go into all of this later. But the plain fact is, and I have been involved in the care of paranoid patients, that they are mistrustful of everyone and getting them to take medication is often difficult because their delusional belief is that the medication is or may be poison. Again Lee was all too trusting, and took the medication when he actually shouldn't have. It was not prescribed for him. The other point that has to be made is that Bruce Lee was on the way up in terms of international stardom. If the Triads had considered pressing him for money, in whatever way, they would want him around as long as possible because Bruce Lee was going to be making a lot of money.

Let's have a brief look at the history of this group, the Triads. In the 1760s, the Heaven and Earth Society, a fraternal organization was founded, and as the society's influence spread throughout

China, it branched into several smaller groups with different names, one of which was the Three Harmonies Society. These societies adopted the triangle as their emblem, usually accompanied by decorative images of swords or portraits of Guan Yu. Guan Yu (Kuan Yu) (died 219) was a general serving under the warlord Liu Bei during the late Eastern Han Dynasty of China. He played a significant role in the civil war that led to the collapse of the Han Dynasty and the establishment of the state of Shu Han during the Three Kingdoms period, of which Liu Bei was the first emperor. Statues used by triads tend to hold the halberd in the left hand, and statues in police stations tend to hold the halberd in the right hand. This signifies which side Guan Yu is worshiped, by the righteous people or vice versa. The appearance of Guan Yu's face for the triads is usually more stern and threatening than the usual statue. This exemplifies the Chinese belief that a code of honor, epitomized by Guan Yu, exists even in the underworld.[3]

As one of the best known Chinese historical figures throughout East Asia, Guan's true life stories, just as in the case of Bruce Lee, have given way to fictionalized ones. For Guan these are mostly found in the historical novel 'Romance of the Three Kingdoms' or passed down the generations, in which his deeds and moral qualities have been lionized. Guan is respected as an epitome of loyalty and righteousness. The term 'triad' was first coined by British authorities in colonial Hong Kong, as a reference to the triads' use of triangular imagery. While never proven, it is 'highly probable' that triad organizations either took after or were originally part of the revolutionary movement called the White Lotus Society, and quite possibly, The Boxers. The Boxer Rebellion, also known as Boxer Uprising or Yihetuan Movement, was a proto-nationalist movement by the Righteous Harmony Society in China between 1898 and 1901, opposing foreign imperialism and Christianity.[4] The Boxers believed that through training, diet, martial arts, and prayer they could perform extraordinary feats, such as flight. Further, they popularly claimed that millions of spirit soldiers would descend from the heavens and assist them in purifying China of foreign influences. The Boxers consisted of local farmers/peasants and other workers who were made desperate by disastrous floods and widespread opium addiction and laid the blame on Christian missionaries, Chinese Christians, and the Europeans colonizing their country. Missionaries were protected under the policy of extraterritoriality. The Boxers called foreigners *"Guizi"* (literally: demons), a deprecatory term, and condemned Chinese Christian converts and Chinese working for foreigners. The Boxers in battle were only lightly armed with rifles and swords, claiming supernatural invulnerability towards blows of cannon, rifle gunshots, and knife attacks. Perhaps it is some of this mythos that has attached itself to Bruce Lee.

Bruce Lee would have known all too well who the triads were. And he would have had to have watched his back, and walked a careful line because they had influence and they had a lot of control. And so the *modus operandi* of their group, where they rarely failed to make a point of what they did as a warning to others, did not make it in any way likely that Bruce Lee was killed by the Triads. He did not die a violent death and there was absolutely no evidence, according to the toxicological reports subsequent to his death that he was poisoned either. Now let's turn to the other alleged means of his murder. The Hong Kong police, as far as I am aware, never opened a case investigating the murder of Bruce Lee and his family never pressed for one. We will discuss this point in more detail later. Let's now examine the claim that he was killed by *dim mak*.

The 'Death-Point Striking' is ingrained in many styles of kung fu and refers to any martial arts technique that can kill using seemingly less than lethal force targeted at specific areas of the body. The concept known as *dim mak* (literally 'press artery'), traces its history to traditional Chinese acupuncture. Tales of its use are often found in the Wuxia genre of Chinese martial arts fiction. *Dim mak* is depicted as a secret body of knowledge with techniques that attack pressure points and meridians of the body, and this is said to incapacitate, or sometimes cause immediate or even delayed death to an opponent. There is no scientific or historical evidence for the existence of a touch of death There have been a number of martial artists claiming to practice the technique in reality, beginning in the

1960s with American eccentric Count Dante, who gave it the English name 'The Death Touch'. However, the subject of the death touch in real life is in much debate and controversy. Despite demonstrations where a subject will seemingly be knocked unconscious, many believe they are actors and are only feigning the effects *dim mak* supposedly has on a person.

Commotio Cordis

As many would know there is a possible medically related condition called *commotio cordis*. *Commotio cordis* is Latin for 'agitation of the heart' and is reported as the second most common cause of sudden cardiac arrest in young athletes, after hypertrophic cardiomyopathy, in the United States. It is seen where fit, strong athletes suddenly collapse after receiving what appears to be an innocuous blow to the chest.[5] Something that they would ordinarily have been able to withstand. *Commotio cordis* is defined as an instantaneous cardiac arrest produced by a nonpenetrating blow to the chest occurring within a specific 10- to 30 millisecond portion of the cardiac cycle in the absence of preexisting heart disease or identifiable morphologic injury to the sternum, ribs, chest wall, or heart. This period occurs in the ascending phase of the T wave, when the ventricular myocardium is repolarizing. With the average cardiac cycle duration of 1000 milliseconds, the probability of a mechanical trauma within the window of vulnerability is only 1-3%. The person who is struck collapses immediately in most instances. In the remaining cases, the individual has a transient period of consciousness, during which a brief purposeful activity, movement, or behavior (e.g., picking up and throwing a ball, crying) is performed before final collapse. According to data collected by the National Commotio Cordis Registry, at the time of the incident, 50% of persons struck were engaged in organized competitive athletics. The remainder were involved in normal daily activities (25%) or recreational sports (25%). Baseball, softball, hockey, and football are the sports most commonly involved. Other associated organized activities included soccer, lacrosse, boxing, and karate. Cases involved with daily activities have included playful boxing, a 'remedy' for hiccups, parental discipline, being struck by a snowball, and an accidental kick during cheerleading, among others. In most instances (58%), the person was struck by a projectile, which was most commonly a pitched, thrown, or batted baseball or softball estimated to be traveling 30-50 miles per hour at most. Other projectiles have included hockey pucks and lacrosse balls. In 42%, chest trauma resulted from bodily contact with another person or a stationary object. Examples of this have included a player's helmet during a football tackle, the heel of a hockey stick, a karate kick, and a body collision. Survival after a *commotio cordis* event is still the exception. Although efforts at resuscitation occur frequently, often involving trained bystanders or emergency medical technicians, the onset of cardiopulmonary resuscitation (CPR) is often delayed because observers underestimate the severity of the trauma or believe that the wind has been knocked out of the person. Survival has usually been associated with effective CPR efforts and defibrillation that occur within 1-3 minutes of the collapse. The survival rate was only 3% in cases in which resuscitative efforts were delayed longer than 3 minutes. Although numerous individuals have been resuscitated with the restoration of a perfusing (effective) heart rhythm, many of these individuals have experienced irreversible ischemic encephalopathy (damage to cells in the brain and spinal cord, from inadequate oxygen) and ultimately died as a result of the injury. Persons with *commotio cordis* are typically found to be unresponsive, apneic, (not breathing) pulseless, and without an audible heartbeat; many are cyanotic, that is they have bluish discoloration of the skin and mucous membranes due to a lack of oxygen in the blood. Grand mal (tonic-clonic) seizures have been evident in some persons with *commotio cordis*. Chest wall contusions and localized bruising that correspond to the site of chest impact are noted over the precordium, that is, the portion of the body over the heart and lower chest, in approximately one third of patients. The ribs and sternum are not structurally injured.[6]

For those of you who are interested in medicine, the clinical aspects of sports injuries, or just generally in such matters we will go a little deeper, because there has been a lot of research done on this condition and anyone taking part in competitive sport should have an awareness of it.

Research has indicated that the location of the blow is critical and it must be directly over the heart, particularly at or near the center of the cardiac silhouette or imagined outline of the heart when looking at the chest. This finding is consistent with clinical observations that precordial bruises representing the imprint of a blow are frequently evident in victims. There is no evidence in humans or in experimental models that blows sustained outside the precordium (e.g., the back, the flank, or the right side of the chest) cause sudden death. The second determinant concerns the timing of the blow, which must occur within a narrow window of 10 to 20 milliseconds on the upstroke of the T wave, just before its peak and accounting for only 1% of the cardiac cycle. That is, the blow must occur during an electrically vulnerable period, when inhomogeneous dispersion of repolarization is greatest, creating a susceptible myocardial substrate for provoked ventricular fibrillation. Don't worry, everyone who is not a cardiologist would have difficulty with the last sentence. It simply refers to the electrical activity of the heart and how at one point in this cycle there is a period of vulnerability in which the regular rhythm of the heart may be 'derailed' into one type of, usually fatal, arrhythmia called ventricular fibrillation. In pigs, on which experiments were conducted, when blows occurred outside this brief window of time, ventricular fibrillation was not the consequence; instead, what followed was transient complete heart block, left bundle-branch block, or ST-segment elevation. These are other forms of heart arrhythmias. These effects have also been reported in some human survivors. Such conditions as these are highly significant in court cases where someone had died after an apparently minor altercation. The accused may claim that they only pushed the person hard in the chest or actually hit them, but not particularly hard. The key determinant would be the exact position of the strike and the precise point in the heart's electrical cycle of activity when the blow occurred. Other factors that may increase the risk of ventricular fibrillation and *commotio cordis* include the hardness of the object and its size and shape, with hard, small, sphere-shaped projectiles most likely to do harm. Interestingly, in regards to our discussion here, adults probably gain a measure of protection from their mature and fully developed chest cage, which may explain in part the apparently low rate of *commotio cordis* events in sports such as kickboxing and boxing, accounting for less than 5% of registry cases. In boxing, it is also possible that the glove itself, which increases the area of impact, helps to buffer the force of the blow. The same type of circumstance can occur with the next 'anatomical vulnerable spot' that we will consider, and which can lead to sudden death.[7]

The Carotid Sinus.

The carotid sinus is another anatomical spot that if struck or stimulated can result in sudden death. The carotid sinus is in the internal carotid artery, immediately after the division of the common carotid artery into the external and internal carotid arteries. In some people the sinus extends into the common carotid artery. Both the right and left internal carotid arteries have a carotid sinus. The sinus is formed by a dilation of the lumen of the artery. In terms of external anatomy, the carotid sinus is located in the neck on either side at about the level of your larynx, that is, the voice box.[8]

'The history of forensic medicine is replete with matters relating to the carotid sinus and the debate about whether in fact it plays a role in sudden death. Pressure on the baroreceptors situated in the carotid sinuses, the carotid sheaths and the carotid body, can result in bradycardia, which is a slowing of the heart, or in total cardiac arrest. It is often claimed, admittedly without much concrete evidence, that fear, apprehension, struggling and possibly the effect of drugs such as alcohol, may heighten the sensitivity of this vagal mechanism. The vagus nerve, also called the pneumogastric nerve or cranial nerve X, is the tenth of twelve (excluding CN0) paired cranial nerves. Parasympathetic innervation of the heart is controlled by the vagus nerve. To be specific, the vagus nerve acts to lower the heart rate. The right vagus innervates the sinoatrial node of the heart. Parasympathetic hyperstimulation predisposes those affected to bradyarrhythmias. The left vagus when hyperstimulated predisposes the heart to atrioventricular (AV) blocks. What this essentially means is that stimulation of this important nerve can result in slowing of the heart and can result in specific cardiac arrhythmias. The release of catecholamines during such adrenal responses may well sensitize the myocardium to such neurogenic stimulation. Catecholamines are hormones produced by the adrenal glands, which are found on top of the kidneys. They are released into the blood during times of physical or emotional stress[9] The vagal reflex has profound implications in relation to pressure or blows on the neck. Sometimes called 'vagal inhibition', 'vasovagal shock' or 'reflex cardiac arrest', the rapid onset of heart stoppage may precede any evidence of congestive or 'asphyxial' (e.g. suffocation or strangulation) signs, causing death immediately or within seconds, or at any time thereafter. It is a matter of some dispute as to whether this reflex can cause immediate cardiac arrest or whether there has to be a period of marked slowing of the heart with negligible cardiac output. Or whether an arrhythmia such as ventricular fibrillation precedes such an arrest. Probably any combination can occur, but it is an indisputable fact that collapse and apparent death can occur immediately on the application of pressure to the neck.[10]

Another cause for sudden cardiac arrest is a blow to the neck or throat. This is the basis of the so-called 'commando punch' and some of the oriental martial arts also contain this in their repertoire, and often forbidden because of its potential lethality. The edge of the hand is brought forcibly across the side of the neck or the front of the larynx. Direct violence to the carotid region naturally causes gross stimulation of the afferent nerve endings. Blows directly to the larynx indirectly stimulate the sinus region or the laryngeal sensory nerve endings and may themselves trigger the cardio-inhibitory reflex. It is well known that the hypopharynx and larynx are particularly sensitive to stimulation, which accounts for the sudden deaths from impaction of food in the larynx, or from the flooding with cold water that causes some sudden immersion deaths. The testicles and uterine cervix also have a similar reputation for leading to sudden cardiac death, if unexpectedly overstimulated, especially when the myocardium is pre-sensitized by catecholamines released by fear or emotion.'[11]

As a matter of interest the above was referenced to one of the forensic works of Professor Bernard Knight. Professor Bernard Knight was born in 1931and became Home Office pathologist in the United Kingdom in 1965. He was appointed Professor of Forensic Pathology, University of Wales College of Medicine, in 1980. He was awarded the CBE (Commander of the Most Excellent Order of the British Empire) in 1993 for services to forensic medicine. As well as being a forensic scientist, and a doctor, he also qualified as a barrister. He was a colleague of another professor of forensic medicine whom we will read about later in this book, Professor Donald Teare. This is the professor who was brought in to the inquest in the death of Bruce Lee to provide specialist forensic opinion, specifically in regards to whether or not marijuana was responsible for Lee's death. Tom Bleecker, in his book 'Unsettled Matters', does not seem to have much time or regard for the determinations of the incredibly experienced and knowledgeable Professor Teare, as we shall see later. However as we know, sometimes experts can get things wrong, there is no such thing as infallibility, and so some cynicism towards experts is healthy. But my own research leads me to the same conclusions that the professor made. As an aside, Professor Teare and Professor Knight collaborated on an early influential study on cot deaths. My research for this book took me, of all places, to an online copy of 'The Glasgow Herald' dated Wednesday, December 23, 1970. The article, which interestingly gives the name of the professor as Dr. R. Donald Teare, which indicates he had a preference for using his middle Christian name over that of Robert, is about cot deaths, which was poorly understood at the time, and still is. No doubt both of these highly trained, experienced and knowledgeable men would have been called to testify in cases where babies that were apparently healthy and full of life had been put in their cots and later were found to be dead. There would have been a high index of suspicion that the baby had been a victim of infanticide. That is, the mother or father had murdered the baby. The job of the professors was to testify that this was not the case, that healthy babies could be victims of sudden death, which had nothing to do with the way they were being cared for by their parents. Cases like these arouse the same kind of incredulity that the death of Bruce Lee raised. But the underlying mechanisms of natural but unexpected death are still at play, although of a different kind. Professor Teare was a man with enormous experience in forensic matters and in medical syndromes related to sudden death, and whether these were was natural or due to a criminal act. We'll deal with exactly how Bruce Lee died later in the book. Now back to our discussion of the various causes of sudden death.

It has been recognized for a considerable time that a 'nervous stimulus' can cause rapid death, and the forensic medico Dr. Casper stated that death from suffocation could happen rapidly from 'nervous apoplexy' in which there was no trace discoverable by the anatomist's knife, that is nothing evident at autopsy so explain what had happened.[12] This form of reflex, commonly called 'vagal inhibition', which has been superseded by the more modern term 'reflex cardiac arrest', is probably initiated by alteration in carotid sinus pressure, the vagus nerve trunk in man being comparatively inexcitable to mechanical stimulation. Carotid sinus stimulation in man can produce effects varying from transient cardiac slowing, hypotension (low blood pressure) and vasodilation, to syncope, cardiac arrest and possibly death.[13] [14] Vasodilation is widening of the blood vessels and syncope is another word for fainting. It is older people who may be more prone to these kinds of events. In older people with advanced arteriosclerosis of the carotid arteries and in rare instances of hypersensitivity of the carotid sinus, a fall or blow on the side of the neck may cause fainting and even cardiac arrest. Again, as it relates to the courts defense attorneys often argue that the death of the victim was unpredictable, without malice or foresight on the part of their client and consequently accidental, in cases of obvious manual strangulation. The following must be considered in determining whether a death may have been caused by carotid sinus reflex:

Deaths as a result of carotid sinus reflex are highly infrequent.

The victim is likely to be elderly with arteriosclerosis involving the carotid arteries.

The victim was known to be susceptible to carotid sinus stimulation with manifestations of dizziness, headaches, weakness and fainting associated with pressure on either side of the neck.

The presence of injuries, especially fingernail marks (whether the victim's or the assailant's) on the skin of the neck, usually rules out death by carotid sinus stimulation.

Injuries in the soft tissues under the skin should be limited to the areas of the carotid sinus.

Absence of petechial hemorrhages in the conjunctivae and internal organs. Petechiae are small (1-2mm) red or purple spots on the body, caused by minor hemorrhaging (broken capillary blood vessels).

What the above means is that basically there *are* various ways that sudden death can be induced, however the important point is that the mechanisms that we have described above relate to sudden death, and sudden death due to mechanisms affecting the cardiovascular system and the central nervous system. As must be repeated Bruce Lee died of cerebral edema. His heart and his breathing ceased as a result of this, due to the pressure within his cranium increasing and mechanically shifting and increasing pressure on delicate and vital brain regions. The principal danger and result of such conditions as cerebral edema is cerebral herniation and there are 3 different types of this. They can occur alone or in combination and they are: *Subfalcial herniation, uncal (transtentorial) herniation,* and *cerebellar tonsillar herniation.* In cerebellar tonsillar herniation there may be stiffness of the neck and head tilt. Compression of the *pons* and *medulla* damages vital centers for respiration and cardiac function, resulting in cardiorespiratory arrest. Pressure on the *posterior fossa* contents from above or from within compresses the *pons* against the *clivus* and displaces the *cerebellar tonsils* into the *foramen magnum*.[15] The cerebral edema developed, in the opinion of the pathologist who conducted the autopsy on Lee, rapidly. So we know that the cardiorespiratory arrest that Bruce Lee suffered was secondary to the process of cerebral edema and presumably brain structure compression and herniation. It was *not* due to any assault or injury or attack on his vital centers directly such as through pressure on the carotid sinus or any such mechanism. If this had occurred there would *not* have been brain edema and his death would have been instantaneous, we can presume. Further, there is absolutely no evidence whatsoever that any of these vital nervous centers, such as the carotid sinus, can be subject to any delayed death effect. It has been postulated and stated by the conspiracy theorists that some weeks prior to his death he was given the 'death touch', or a kung fu practitioner used the 'quivering palm technique' against him and this led to his demise.[16] The probability of this providing a realistic scenario and reason for his death is about the same probability that Santa Claus exists and that we are visited by tooth fairies. It's gotten to the point where some martial arts practitioners have claimed that they can utilize 'qi-based' attacks that work without actual physical contact. I won't even talk about probabilities here. Such claims are just plain silly. It's the same kind of claims that are made for so-called 'psychic surgery' in the country in which I live, the Philippines, which I have researched and examine in my book 'Death and Afterlife: The Philippine Experience.' Under proper scientific scrutiny, and rational research, all such claims evaporate.

Poisoning

In regards to the claim that Bruce Lee was poisoned let's have a look at this in a little more detail. There was no evidence of this during the post mortem examination as far as I can determine from the information that we have which indicates possible poisoning, and the toxicological results were all negative. Cases of poisoning will in most cases exhibit clear post mortem appearances in the victim. In many poisons and classes of poisons they are characteristic and unmistakable. Chemical (toxicological) analysis will take place and this is one of the most important forms of evidence, as a demonstration of the actual presence of a poison in the body carries immense weight. The poison may be discovered in the living person by testing the urine, the blood abstracted by bleeding, or the serum of a blister. In the dead body it may be found in the blood, muscles, and viscera and especially in the liver, and secretions. Its discovery in these cases must be taken as conclusive evidence of administration. If, however, it be found only in substances rejected or voided from the body, the evidence is not so conclusive, as it may be contended that the poison was introduced into or formed in the material examined after its rejection from the body, or if the quantity be very minute it will be argued that it is not sufficient to cause death. Whilst recognizing the fact that toxic agents cannot be accurately classified, the following grouping may for descriptive purposes:

Corrosives are characterized by their destructive action on tissues with which they come in contact. The principal inorganic corrosives are the mineral acids, the caustic alkalies, and their carbonates; the organic are carbolic acid, strong solutions of oxalic acid, and acetic acid.

The symptoms of poisoning by corrosives include: Burning pain in mouth, throat, and gullet, strong acid, metallic or alkaline taste; retching and vomiting, the discharged matters containing shreds of mucus, blood, and the lining membrane of the passages. Inside of mouth corroded. There is also dysphagia, thirst, dyspnoea, a rapid and feeble pulse, anxious expression, shock. Death may result from shock, destruction of the parts, e.g. perforation of stomach or duodenum, suffocation; or some weeks subsequently death may be due to cicatricial contraction of the gullet, stomach, or pylorus. (The word cicatricial had me running to my medical dictionary too. This simply refers to a scar left by the formation of new connective tissue over a healing sore or wound.)

The post mortem appearances or signs of corrosives poisoning are distinctive and include corrugation from strong contraction of muscular fibers, and followed by inflammation and its consequences. The mouth, gullet, and stomach, and in some cases the intestines, may be white, yellow, or brown, shriveled and corroded. The corrosions may be small, or may extend over a very large surface. Sometimes considerable portions of the lining membrane of the gullet or the stomach may be discharged by vomiting or by stool. Beyond the corroded parts the textures are acutely inflamed. The stomach is filled with a yellow, brown, or black gelatinous liquid or black blood, and may in rare cases be perforated. It is not a pretty sight at all and not a pleasant way to die.

Irritants are the next major group we will consider and these are substances which inflame parts to which they are applied. The class includes mineral, animal, and vegetable substances, and contains a larger number of poisons than all the other classes together. Irritants may be divided into two groups: (1) Those which destroy life by the irritation they set up in the parts to which they are applied; (2) those which add to local irritation peculiar or specific remote effects. The first group includes the principal vegetable irritants, some alkaline salts, some metallic poisons for instance and the second comprises the metallic irritants, the metalloids (phosphorus and iodine), and one animal substance, cantharides.

The symptoms of poisoning by irritants include burning pain and constriction in throat and gullet, pain and tenderness of stomach and bowels, intense thirst, nausea, vomiting, purging and tenesmus, with bloody stools, dysuria, cold skin, and feeble and irregular pulse.(Tenesmus is the feeling that you constantly need to pass stools, even though your bowels are already empty.)The vomit consists at first of food, and then it becomes bile-stained and later dark coffee-grounds in appearance, due to extravasation of blood from the over-distended vessels in the gastric mucous membrane. Death may occur from shock, convulsions, collapse, exhaustion, or from starvation on account of chronic inflammation of the gastro-intestinal mucous membrane.

The post mortem appearances in the body are those of inflammation and its consequences. The surface of stomach, fauces, gullet, and duodenum, may be thickened, black, ulcerated, gangrenous, or sloughing. Vessels filled with dark blood ramify over the surface. Acute inflammation is often found in the small intestines, with ulceration and softening of mucous membrane. The rectum is frequently the seat of marked ulceration. It is readily apparent to any pathologist when this type of poisoning has occurred.

The next general group of poisons is those acting on the brain. Three further classes or subgroups are usually described in the literature: The opium group, producing sleep; the belladonna group, producing delirium and illusions; and the alcohol group, causing exhilaration, followed by delirium or sleep.

In regards to symptoms those of the opium group include giddiness, headache, dimness of sight, contraction of the pupils, noises in the ears, drowsiness and confusion, passing into insensibility. Of the belladonna group, delirium, illusions regarding sight, (misinterpreting visual stimuli) dilated pupils, dry mouth, thirst, redness of skin, coma. Of the alcohol group, excitement of circulation and of cerebral functions, want of power of co-ordination and of muscular movement, double vision, mania, followed by profound sleep and coma. In the chronic form, delirium tremens.

The post mortem appearances in the opium group include fullness of the sinuses and veins of the brain, with effusion of serum into the ventricles and beneath the membranes. In regards to the belladonna group I found reference to this in an old text:

'Although as I have said convulsions are rare yet lock jaw *subsultus tendinum* and occasionally chorea (irregular, rapid, uncontrolled, involuntary, excessive movement that seems to move randomly from one part of the body to another - author) or at least symptoms closely resembling it have occurred. Strangnary (this is an old term for difficulty in passing urine - author) and suppression of urine have also occurred. In the post mortem examination of the body the most striking circumstance is the rapid decomposition of the body putrefaction begins in less than twelve hours after death. The blood vessels of the head are gorged and the stomach and intestines show the marks of inflammation. The best mode of averting the fatal effects of this poison is to exhibit emetics of sulphate of zinc or of copper assisting their operation by irritation of the fauces then evacuating the bowels by active purgatives and glysters and following these by large doses of vinegar or other vegetable acids.[17] The text that I sourced this from is a very old edition of the British medical journal The Lancet, but it is nonetheless relevant today.

In the alcohol group, signs of inflammation, congestion of brain and membranes, fluidity of blood and long-continued rigor mortis of the corpse.

In the fourth group are included poisons acting on the spinal cord, and these include strychnine, brucine and thebaïne. The leading symptom is tetanic spasm, which is a sudden, violent, involuntary muscular contraction.

Poisons affecting the heart kill by sudden shock, syncope, or collapse. They comprise prussic acid, dilute solution of oxalic acid and oxalates, aconite, digitalis, strophanthus, convallaria, and tobacco. Then there are poisons acting on the lungs. These have for their type carbonic acid gas and coal gas. The fumes of ammonia are intensely irritating, and may give rise to laryngitis, bronchitis, and even pneumonia. Nitric acid fumes sometimes produce no serious symptoms for an hour or more, but there may then be coughing, difficulty of breathing and tightness in the lower part of the throat, followed by capillary bronchitis.[18] All of these poisons above and any symptoms or post mortem signs of their presence or effects would be apparent to any pathologist.

So rule out *dim mak* and murder by the Triads as being the cause of Bruce Lee's death. There is absolutely no evidence for either. But some stubbornly stick to it. In an interview that Tom Bleecker gave to an internet site by the name of Divine Wind that I read as part of my research I came across the following exchange:

'DW Forum - I can't imagine anyone ever owning up to taking part in a cover up, so what evidence would be needed to open an official investigation into the deaths of Bruce and Brandon ?

Tom Bleecker - The issue of the death of Bruce Lee is closed forever. The reason being that the English Hong Kong police department and British government no longer exists. Hong Kong has now returned to Mainland China.'[19]

Those who believe that Lee was poisoned or murdered have repeated this view, that it would not be possible to re-open any investigation, that the communists now in power in Hong Kong would not allow an investigation or trial to take place. There is only one problem with their stance on this as I see it. Bruce Lee died in 1973 and Hong Kong was handed over to the communists in 1997. That's a full 24 years that these individuals had to petition the British authorities to re-open the case, or to charge someone with murder. And they didn't. What were all these individuals doing for the all that time prior to the hand over of Hong Kong to China? Reportedly some of Lee's family suspected foul play. Even if the Hong Kong police did stamp his death initially as 'probable homicide' they would probably do exactly the same for any other 32 year old male like Lee who died suddenly. Especially with the presence of the Triads in Hong Kong. There would be a high index of suspicion for homicide, however the subsequent inquest and autopsy revealed no evidence of any nature that would stand up to scrutiny in court in regards to any such allegation.

The interview with Tom Bleecker that I cited above continues:

'DW Forum - Was Bruce Lee beaten up on July 20th 1973, no way medicine can do that to your face. He was poisoned, went to bed, then beaten up, then they put his clothes on and forget about his second show?

Tom Bleecker - I seriously doubt that Bruce as beaten up on July 20th. The swelling of his face is not uncommon. Compare Bruce's face to that of Elvis Presley. The swelling was caused by retention of water, in his face and his brain.'[20]

Tom Bleecker appropriately dismisses the suggestion that Bruce Lee was beaten up. He may have had swelling to his face, and there are two possible reasons for this. One is that the swelling is angioedema (facial swelling) which took place as part of an anaphylactic reaction which caused Lee's cerebral edema after he took the Equagesic tablet. Another more mundane reason may be that those who were present with him on the night of his death, when he lapsed into a coma, instead of calling for an ambulance reportedly kept slapping his face to attempt to wake him up. It could even be a combination of both of these reasons.

Rumors and theories about his death being the result of murder or poisoning have been going on since his passing, without the slightest piece of hard evidence to back them up. Only speculation and various scenarios based on assumption and conjecture. Bruce Lee as a martial artist stressed simplicity and directness in all things. However such theories exist because conspiracy theory is the currency of the 21st Century, and there is a huge book and movie industry that surrounds such matters. The Roswell incident, 9/11 being a conspiracy manufactured by intelligence agencies, and the claim that there was never a moon landing by Apollo astronauts are examples that come to mind. People may enjoy reading them, but as long as they understand they do not equate with the truth then we can live with the nonsense that is out there. Where is my evidence that Bruce Lee was not murdered or poisoned? I don't need any! *The onus is on those who make the claim that he allegedly was poisoned to provide the proof.* And they have failed spectacularly to provide any such proof or objective evidence that would stand up in a court of law. And they did nothing to petition the authorities in Hong Kong to have charges brought against someone with murder for 24 years, as far as I am aware. Bruce Lee was not murdered, he was not poisoned, and he did not die as the result of a kung fu 'death touch'. Sorry to disappoint, but the world and the events in it mostly occur for simple and mundane reasons. Even the tragic deaths of 32 year old kung fu movie megastars, such as that of Bruce Lee.

Chapter 2
Cerebral Edema

As I have stated above the fact is that Bruce Lee died as a result of cerebral edema. Let's look at this condition and obtain clinical details in regards to what it actually is, what its effects are, and what the causes of it are.

Cerebral edema, or brain swelling, can be caused by over 300 separate conditions and diseases.[1] This is very important to remember, in view of our discussion of the death of Bruce Lee. It is entirely possible to experience cerebral edema at one specific point in time for one specific reason or cause, and to further suffer it at a later date, for an entirely different reason. Brain edema can best be defined as an increase in brain volume. One of the key issues to remember about the brain is that it is enclosed in a non-expandable case: the cranium. When most other organs enlarge, there is generally some room for expansion. The brain, when it expands due to a mass lesion (such as a tumor or a hemorrhage) or generalized edema, it has very little space in which to expand. Not only does the mass cause direct tissue damage, but it may also cause problems by initiating adjacent cerebral edema and increasing the intracranial pressure. If intracranial pressure increases enough, it will decrease blood flow to the brain and cause characteristic clinical symptoms.[2] This we have discussed above in regards to cerebral herniation syndromes. In cerebral edema this increase in intra- or extracellular volume can be from an excess in either osmotic ally active particles or water. A noncontract CT scan of the head suggestive of cerebral edema, which is a diagnostic method that was unavailable in the days when Bruce Lee died, will show loss of grey-white matter differentiation highlighting the basal ganglia and cortical grey matter with low density noted in the white matter. White matter tissue (composed primarily of axons, myelin, and type 2 atrocities) has been identified as more susceptible to edema formation than grey matter tissue (composed primarily of neurons and type 1 atrocities). Recent work has helped to delineate the increased susceptibility of atrocities to swell. Atrocities are large, star-shaped cells that hold nerve cells in place and help them develop and work the way they should. They are intimately involved in controlling the flux of water across the blood-brain barrier, which is composed primarily of atrocities and astrocytic foot processes interfacing with the tight junctions between endothelial cells.[3] Cerebral edema is frequently encountered in clinical practice from diverse origins and is a major cause of increased morbidity and death in this subset of patients. *The consequences of cerebral edema can be lethal and include cerebral ischemia from compromised regional or global cerebral blood flow (CBF) and intracranial compartmental shifts due to intracranial pressure gradients that result in compression of vital brain structures.* (Author's italics) The overall goal of medical management of this condition is to maintain regional and global cerebral blood flow to meet the metabolic requirements of the brain and *prevent secondary neuronal injury from cerebral ischemia.* (Author's italics) Medical management of cerebral edema involves using a systematic and algorithmic approach. This would include general measures, such as optimal head and neck positioning for facilitating intracranial venous outflow, avoidance of dehydration and systemic hypotension, and maintenance of normothermia, that is, normal body temperature. (Refer to the May 10, 1973 episode of cerebral edema when Lee was reported to have had a 'high fever'.) Further specific therapeutic interventions should occur including controlled hyperventilation, administration of corticosteroids and diuretics, osmotherapy (such as Mannitol osmotherapy – author), and pharmacological cerebral metabolic suppression.[4] And so cerebral edema is a nasty clinical syndrome that can easily lead to death if it is not managed correctly or proper medical intervention occurs too late, and from which, even if the person survives, there may be secondary and lasting damage to the brain due to cerebral ischemia where blood flow has been compromised and neurons have died.

Further to the more than 300 different conditions that can cause cerebral edema the condition itself is further classified into four types and they are:

Vasogenic

This occurs due to a breakdown of tight endothelial junctions which make up the blood–brain barrier (BBB) (The blood–brain barrier acts very effectively to protect the brain from many common bacterial infections.) This allows normally excluded intravascular proteins and fluid to penetrate into cerebral parenchymal extracellular space. Once plasma constituents cross the BBB, the edema spreads; this may be quite fast and widespread. As water enters white matter it moves extracellularly along fiber tracts and can also affect the gray matter. This type of edema is seen in response to trauma, tumors, focal inflammation, late stages of cerebral ischemia and hypertensive encephalopathy. Some of the mechanisms contributing to BBB dysfunction are: physical disruption by arterial hypertension or trauma, tumor-facilitated release of vasoactive and endothelial destructive compounds. For example, arachidonic acid, excitatory neurotransmitters, eicosanoids, bradykinin, histamine, (*a substance that mediates the allergic response* – author) and free radicals.

There are further sub-categories of vasogenic cerebral edema and these include:

Hydrostatic cerebral edema.

This form of cerebral edema is seen in acute, malignant hypertension. It is thought to result from direct transmission of pressure to cerebral capillary with transudation of fluid into the ECF (extracellular fluid) from the capillaries.

Cerebral edema from brain cancer.

Cancerous glial cells (glioma) of the brain can increase secretion of vascular endothelial growth factor (VEGF) which weakens the junctions of the blood–brain barrier. Dexamethasone can be of benefit in reducing VEGF secretion.

High altitude cerebral edema.

High altitude cerebral edema (or HACE) is a severe form of (sometimes fatal) altitude sickness. HACE is the result of swelling of brain tissue from leakage of fluids from the capillaries due to the effects of hypoxia on the mitochondria-rich endothelial cells of the blood–brain barrier. Symptoms of this can include headache, loss of coordination (ataxia), weakness, and decreasing levels of consciousness including disorientation, loss of memory, hallucinations, psychotic behavior, and coma. It generally occurs after a week or more at high altitude. Severe instances can lead to death if not treated quickly. Immediate descent is a necessary life-saving measure (2,000 - 4,000 feet). There are some medications such as dexamethasone that may be prescribed for treatment in the field, but these require proper medical training in their use. Anyone suffering from HACE must be evacuated to a medical facility for proper follow-up treatment. Climbers may also suffer high altitude pulmonary edema (HAPE), which affects the lungs. While not as life threatening as HACE in the initial stages, failure to descend to lower altitudes or receive medical treatment can also lead to death.

Cytotoxic

In this type of edema the BBB remains intact. This edema is due to the derangement in cellular metabolism resulting in inadequate functioning of the sodium and potassium pump in the glial cell membrane. As a result there is cellular retention of sodium and water. There are swollen astrocytes in gray and white matter. Cytotoxic edema is seen with various intoxications such as that caused by dinitrophenol, triethyltin, hexachlorophene or isoniazid, in Reye's syndrome, severe hypothermia, early ischemia, encephalopathy, early stroke or hypoxia, cardiac arrest, pseudotumor cerebri, and cerebral toxins.

Osmotic

Normally cerebral-spinal fluid (CSF) and extracellular fluid (ECF) osmolality, or the concentration of a solution in terms of osmoles of solute per kilogram of solvent, of the brain is slightly lower than that of plasma. When plasma is diluted by excessive water intake (or hyponatremia), syndrome of inappropriate antidiuretic hormone secretion (SIADH), hemodialysis, or rapid reduction of blood glucose in hyperosmolar hyperglycemic state (HHS), formerly hyperosmolar non-ketotic acidosis, the brain osmolality will then exceed the serum osmolality creating an abnormal pressure gradient down which water will flow into the brain causing edema. Mannitol on the other hand is often given to reverse cerebral edema. Mannitol is an osmotically active substance that does not pass into the cell. Instead it draws water from the intracellular space into the extracellular space where it can be removed by the kidneys.

Interstitial

Occurs in obstructive hydrocephalus. This form of edema is due to rupture of the CSF-brain barrier resulting in trans-ependymal flow of CSF which causes CSF to penetrate the brain and spread to the extracellular spaces and the white matter. This is differentiated from vasogenic edema in that interstitial cerebral edema CSF contains almost no protein. Treatment approaches can include, again, osmotherapy using Mannitol, diuretics and surgical decompression.

The signs and symptoms of cerebral edema, and associated raised intracranial pressure, include:

Headache, worsened with Valsalva maneuver (The Valsalva maneuver is performed by attempting to forcibly exhale while keeping the mouth and nose closed.)

Decreased visual acuity (Disturbance of vision)
Diplopia (Double vision)
Nausea
Vomiting
Papilledema (Swelling of the optic disc)
Progressive decline in level of consciousness
Decreased upward gaze
Cranial nerve VI palsy
Loss of normal venous pulsations in the fundus
Visual field cut or enlarged physiologic blind spot
Alterations in vital signs (Heart rate and respiration)
Neck pain or stiffness
Dizziness
Memory loss
Inability to walk
Difficulty speaking
Stupor
Seizures
Loss of consciousness [5]

Bruce Lee was successfully treated, by the use of Mannitol, for his first episode of cerebral edema that occurred on May 10, 1973. The other osmotherapy option is hypertonic saline. There are lingering effects from an episode of cerebral edema that has been successfully medically treated and these can include:

Sleep disturbance
Disruption of thinking and attention skills
Headaches
Depression
Impairment of communication skills
Movement impairment

In regards to the detection and diagnosis of cerebral edema physicians, nurses and paramedics will be familiar with this process, which includes a comprehensive neurologic screening assessment. This can be accomplished within minutes if performed in an organized and systematic manner.[6] Neurologic screening assessment includes six major components of the neurologic examination, namely:

1) Mental status examination (the behavior and thought processes of the individual as well as level of consciousness for instance)
2) Cranial nerve examination
3) Motor examination (Examination for motor dysfunction includes assessment of strength, muscle tone, coordination and any abnormal movements)
4) Reflexes
5) Sensory examination
6) Evaluation of coordination and balance.

Based on the chief findings of the screening assessment, further evaluation or investigations can be then decided upon.[7] Obviously, where symptoms are present indicating cerebral edema the individual should be transported to an emergency department by ambulance and reviewed by a physician *immediately*. It is a medical emergency. Cerebral edema is a challenging problem even in the neurocritical care setting. Secondary brain injury may ensue as a result of cerebral edema, and may result in different herniation syndromes.

Any doctor, paramedic or nurse would know and recognize the above symptoms. We know that Bruce Lee experienced a progressive decline in his level of consciousness until the point where he was comatose. At that point, as stated, a doctor or an ambulance should have been summoned *immediately*, especially with his recent medical history of an episode of cerebral edema and a tonic-clonic (grand-mal) seizure. Medical examination would have then determined the presence of the other above-named physical signs and intervention could have been initiated.

Reye's syndrome may be worth mentioning here. It is a serious medical condition associated with viral infection and aspirin intake. It usually strikes children under age 18, most commonly those between the ages of 5 and 12; however cases in adults have been documented. Symptoms of Reye's syndrome develop after the patient appears to have recovered from the initial viral infection. Symptoms include fatigue, irritability, and severe vomiting. Eventually, neurological symptoms such as delirium and coma may appear. One third of all Reye's syndrome patients die, usually from progressive heart failure, gastrointestinal bleeding, kidney failure, or *cerebral edema*.[8] (Author's italics) From another source we find the following, 'The exact cause of Reye's syndrome is not understood, but researchers feel that it has something to do with a problem involving the energy-converting structures in the cells of the body.' Further, 'Fluid can accumulate in the brain, causing extensive pressure, which in turn constricts blood vessels, preventing blood flow to the brain. Brain damage and even death can be the end result of this process. There have been cases of Reye's syndrome occurring in adults, and there are some examples where aspirin was not involved, but the vast majority of cases are seen when children have been given aspirin to fight an upper-respiratory viral ailment. The aspirin link to Reye's syndrome is not understood, but it exists undeniably. Symptoms of Reye's syndrome begin about three days to a week after a viral infection.... The individual will become unusually tired and sleepy, and can exhibit disoriented and/or combative behavior. These symptoms come on swiftly, and can get worse as the hours pass. The disease will advance to the point where there is weakness or even paralysis in the legs and arms, a loss of consciousness, or convulsions and seizures.[9] And what was one of the ingredients that Lee ingested as part of the Equagesic tablet? Aspirin. As Professor Teare stated at the inquest in Hong Kong, it could have been a hypersensitivity reaction to *either* to meprobamate or aspirin, or a combination of both.[10] I mention conditions such as Reye's syndrome to highlight the fact that seemingly innocuous non-prescription medications such as Aspirin can have fatal side-effects.

The autopsy report in regards to Bruce Lee is reported by Mr. Tom Bleecker in 'Unsettled Matters' who would have had access to the original and full report. The autopsy report is unfortunately not available as far as I can determine to the public or researchers such as myself. It was conducted by a Dr. Lycette and was essentially normal except for the determination of the cause of death being cerebral edema. There was some minor hemorrhaging of the covering of the lungs and moderate congestion, perhaps consistent with someone with a common cold. [11] The heart was normal. Let's just provide some comment here. In a person who died from anaphylaxis, autopsy may show what is referred to as an 'empty heart' attributed to reduced venous return from vasodilation and redistribution of intravascular volume from the central to the peripheral compartment. [12] This can also be explained in the following manner, 'When loss of central blood volume becomes extreme, the mechanosensitive receptors that normally respond to low volumes and pressures send paradoxical signals to the brainstem cardiovascular centers that are misinterpreted and trigger increases in vagal outflow to the heart and reductions in sympathetic outflow to the peripheral circulation. This is the so-called 'empty heart syndrome' which has been used to explain vasovagal syncope and the physiological responses seen during severe hemorrhage. Other signs of anaphylaxis are laryngeal edema, (the vocal cords in the autopsy report were free of swelling) eosinophilia in lungs, heart and tissues, (eosinophilia is increased eosinophilic leukocytes in the blood) and evidence of myocardial hypoperfusion. [13] In many cases the lack of reliable laboratory biomarkers and common standard definitions of signs and symptoms represents the main problem for clinicians when a suspected anaphylactic event must be diagnosed, while a post-mortem diagnosis of anaphylaxis is often a very difficult task in forensic medicine. Significant necroscopic signs as well as the data reported from witnesses or medical records may be absent, biological fluids as blood or urine may be unavailable or under thanatological modifications. (That's a fancy medical of saying that biological fluids will have become altered chemically due to the process of death - author.) The aim of the review which I sourced this information from was to focus on the diagnostic difficulties with which coroners and forensic pathologists have to cope with when a confirmation of anaphylactic death is required by judicial authorities.[14] So, it's not an easy task at all diagnosing an anaphylactic reaction post-mortem. With regards to the issue of the differences between allergy and hypersensitivity some authorities actually use the terms interchangeably. Currently allergy is considered to be synonymous with hypersensitivity in meaning.[15] They usually refer to 'type 1 immediate hypersensitivity', mediated by specific IgE antibodies in genetically predisposed individuals and resulting in symptoms characteristic of eczema, urticaria, rhinitis, asthma and anaphylaxis, although it is noted that several types of allergic states encompass all the mechanisms described by Gell and Coombs.[16] Gell and Coombs were medical research pioneers in this field, but theirs has not been the last word. Some immunologists have called for other classifications systems, stating that, '…. the classical work of Gell and Coombs in classifying hypersensitivity reactions has not stood the test of time very well.'[17]

As an example of the difference between allergy or hypersensitivity to a drug and intolerance to it, reactions to Aspirin can be considered. Intolerance to Aspirin is characterized by hemorrhages in the stomach whereas allergy to aspirin results in such symptoms as hives, asthma, allergic nasal and sinus disease or even anaphylactic shock.[18] And these events can occur even if the individual has taken the medication without incident for some time. Hypersensitivity reactions are of major concern and present a burden for national healthcare systems due to their often severe nature, high rate of hospital admissions and high mortality. The pathophysiological mechanisms underlying hypersensitivity reactions are not well understood, but general agreement among clinicians and researchers is that they are immune mediated. One of the most commonly reported reactions is delayed type hypersensitivity, which is T cell mediated. Delayed type hypersensitivity reactions can be induced by a large number of drugs. They can occur a few hours or up to several weeks after drug intake, and therefore causality is sometimes difficult to establish based on time-event relationships.

And so the autopsy report in its initial stages showed nothing of real significance except for what may have been the symptoms typical of a cold.[19] Viral infections have been suggested as a potential trigger for hypersensitivity reactions. There are also clear genetic and racial factors with some racial groups being more sensitive to certain medications.[20] As has been stated, allergic reactions and hypersensitivity reactions are extremely complex and not fully understood by medical researchers. There is also the issue of idiosyncratic reactions, genetic vulnerabilities and racial factors that have to be considered also. If, by any chance, an immunologist happens to read this book I would be interested in their comments and their criticisms. This is how research is conducted and how knowledge progresses. No one has a monopoly on the truth and often such clinical issues require further review and revision. Just as I am reviewing the works of others, others can criticize this book too. As long as this is done in a rational way, with evidence and referencing of sources, then we can engage in proper debate and discussion and reach valid conclusions.

Let's move on with the rest of the autopsy report. Due to copyright considerations I am limited as to what I can quote directly from 'Unsettled Matters' and again I refer to readers to that book to read it in its entirety. What occurred further during the autopsy is that Dr. Lycette, as is standard practice, sent samples of urine, blood, liver, kidney, and small intestine and stomach contents to laboratories in Australia and New Zealand as well as Hong Kong. Apart from the small amount of marijuana in Lee's stomach and small intestine there was nothing else of significance. Tests for trace metals indicative of poisoning, alcohol, and other drugs were all negative. Lee hadn't even taken the Dilantin that he had been prescribed some weeks earlier after his medical examination in the United States.[21] Dilantin is an anti-convulsant, used to control epilepsy. Let us provide some further comment here. Despite any insinuations and allegations made by anyone that Bruce Lee was some kind of tablet munching, drug abusing, alcohol swilling, out of control junkie, he had consumed on the day of his death, well, *just about next to nothing*. Apart from marijuana there were no other medications or drugs present. And the Equagesic tablet, please remember, was given to him, he himself did not initiate the taking of this medication. And it was given to him with fatal consequences. She could not have predicted the effect the medication would have on Bruce Lee, and neither could anyone else have.

The autopsy proceeded to the point where the cause of death was evident. There was no evidence of any skull injuries but Lee's brain weighed 1,575 grams compared with the normal 1,400 grams. The blood vessels of the brain were all intact. It was clear that Lee had died of cerebral edema, and the Dr. Lycette's view was that this had developed very rapidly.[22]

Mr. Bleecker goes on to write, 'The conclusion that Bruce Lee died from cerebral edema has never been challenged by anyone. The problem has been in accepting that a single tablet of Equagesic is what caused the fatal edema.'[23] For anyone who knows of the toxic effects or adverse reactions that Aspirin or Meprobamate or the two combined can have it is not difficult to accept at all. Over the intervening period since Lee's death reportedly many have raised the question of whether there was a direct relationship between Lee's collapse on May 10, 1973 and his actual death ten weeks later on July 20, 1973. The view that has been expounded by some is that someone, for some motive, had attempted to kill Bruce Lee on May 10, 1973, when he had his first episode of cerebral edema, and having failed, they engaged in another attempt which took place ten weeks later, and this one succeeded.

We have been told that in regards to the May 10 incident that Lee, while in the dubbing room at Golden Harvest, felt nauseated. Excusing himself, he then grew weak, and perhaps fainted, in a nearby restroom. Twenty minutes later he regained his strength and returned to the dubbing room, where he collapsed. Within minutes he vomited and began convulsing. Another twenty minutes elapsed, after which he was taken to Baptist Hospital where he was found to have a very high fever and was having difficulty breathing. Soon the doctors detected brain swelling, which they were able to control by administering the drug Mannitol. An hour and a half later, Bruce revived and, following a period of amnesia, returned to normal.

There may in fact have been some persisting symptoms of cognitive impairment, or memory problems, which are not unusual after such events, as I state above when discussing cerebral edema.

The May 10 event has also been reported in the following way from another source,

'In early 1973, filming began for Bruce Lee's next film, Enter the Dragon. When shooting was complete, the dubbing process began (due to the high volume of street noise in Hong Kong, outdoor dialogue in the movie had to be dubbed subsequent to the on-screen action). On May 10, 1973, Bruce was engaged in one such dubbing session at Golden Harvest. The air-conditioning had been turned off so as to cut down on noise contamination. This caused the temperature to soar in the tiny dubbing room. Bruce, feeling faint, needed a break and stepped out to the lavatory to throw some water on his face. Once in the lavatory, Bruce passed out. After twenty minutes, a studio assistant found Bruce and helped him to his feet. While being helped back to the dubbing room, Bruce collapsed again and was rushed to a nearby hospital. Doctors found him unresponsive and running a fever of 105°. It was discovered that Bruce had a swelling of fluid that was causing pressure on the brain. He was given Mannitol and he eventually regained consciousness. Upon recovery, Bruce and his wife flew back to Los Angeles. Bruce was given a battery of tests at the University of California Medical Center and emerged with a clean bill of health. Doctors agreed that Bruce had suffered cerebral edema and they gave him a prescription for Dilantin, a drug given to epileptics.'[23]

The above report on the May 10 event gives the cause of this episode as '....a swelling of fluid that was causing pressure *on* the brain.' (Author's italics) In 'Unsettled Matters' Tom Bleecker correctly reports, in regards to the autopsy, that Dr. Lycette had noted an absence of any skull injuries but that Lee's brain was swollen like a sponge and weighed 1,575 grams compared with the normal 1,400 grams. That is, that in cerebral edema the swelling occurs *within the brain tissue,* the accumulation of fluid in the brain substance itself and not *on* the brain.[24] There is a completely different medical condition called hydrocephalus, and it is in this condition that there is an increase in the pressure of the cerebrospinal fluid that surrounds the brain and thus places pressure *on* the brain, although to be exact the ventricles within the brain can be affected. But these are separate and distinct medical conditions. They can occur together as in interstitial cerebral edema as we mention above. This form of edema occurs as a result of obstructive hydrocephalus and is due to rupture of the cerebrospinal-brain barrier resulting in trans-ependymal flow of cerebrospinal fluid (CSF) which causes CSF to penetrate the brain and spread to the extracellular spaces and the white matter. Now, there are other conditions that we need to briefly consider, such as idiopathic intracranial hypertension (IIH). IIH can only be diagnosed if there is no alternative explanation for the symptoms. Intracranial pressure may be increased due to medications such as high-dose vitamin A derivatives (e.g. *isotretinoin for acne*), long-term tetracycline antibiotics (for a variety of skin conditions), and hormonal contraceptives. There are numerous other diseases, mostly rare conditions, which may lead to intracranial hypertension. If there is an underlying cause, the condition is termed 'secondary intracranial hypertension'. Common causes of secondary intracranial hypertension include obstructive sleep apnea (a sleep-related breathing disorder), systemic lupus erythematosis (SLE), chronic kidney disease, and Behçet's disease.[25] (Author's italics) The most common symptom of IIH is headache, which occurs in almost all (92–94%) cases. It is characteristically worse in the morning, generalized in character, and throbbing in nature. It may be associated with nausea and vomiting. The headache can be made worse by any activity that further increases the intracranial pressure, such as coughing and sneezing. (We mentioned the valsalva maneuver above which also causes such an increase) The pain may also be experienced in the neck and shoulders. Many have pulsatile tinnitus, a 'whooshing' sensation in one or both ears (64–87%); this sound is synchronous with the pulse. Various other symptoms, such as numbness of the extremities, generalized weakness, loss of smell, and dyscoordination are reported more rarely; none are specific for IIH. The increased pressure leads to compression and traction of the cranial nerves, a group of nerves that arise from the brain stem and supply the face and neck. Most commonly, the abducens nerve (sixth nerve) is involved. This nerve supplies the muscle that pulls the eye outward. Those with sixth nerve palsy therefore experience horizontal double vision which is worse when looking towards the affected side. More rarely, the oculomotor nerve and trochlear nerve (third and fourth nerve palsy, respectively) are affected; both play a role in eye movements. The facial nerve (seventh cranial nerve) is affected occasionally and the result is total or partial weakness of the muscles of facial expression on one or both sides of the face. The increased intracranial pressure leads to papilledema, which is swelling of the optic disc, the spot where the optic nerve enters the eyeball. This occurs in practically all cases of IIH, but not everyone experiences symptoms from this. Those who do experience symptoms typically report 'transient visual obscurations', episodes of difficulty seeing that occur in both eyes but not necessarily at the same time. Long-term untreated papilledema leads to visual loss, initially in the periphery but progressively towards the center of vision.[26] Physical examination of the nervous system is typically normal apart from the presence of papilledema, which is seen on examination of the eye with a small device called an ophthalmoscope or in more detail with a fundus camera. If there are cranial nerve abnormalities, these may be noticed on eye examination in the form of a squint (third, fourth, or sixth nerve palsy) or as facial nerve palsy. If the papilledema has been longstanding, visual fields may be constricted and visual acuity may be decreased. Visual field testing by automated (Humphrey) perimetry is recommended as other methods of testing may be less accurate. Longstanding papilledema leads to optic atrophy, in which the disc looks pale and visual loss tends to be advanced.[27]

So, to recap, surprising as it may sound, cerebral edema is a fairly common pathophysiological entity which is encountered in many clinical conditions. Many of these conditions present as medical and surgical emergencies. By definition cerebral edema is the excess accumulation of water in the intra-and/or extracellular spaces of the brain. The most rapid and effective means of decreasing tissue water and brain bulk is osmotherapy. Osmotic therapy is intended to draw water out of the brain by an osmotic gradient and to decrease blood viscosity. These changes would decrease ICP and increase cerebral blood flow (CBF). Mannitol is the most popular osmotic agent.[28] In fact, it is standard treatment for this condition.

I include the above to illustrate that the literature reports a variety of significant and unmistakable symptoms present in such conditions as idiopathic intracranial hypertension and cerebral edema. Betty Ting reported that on the night Bruce Lee died he was only complaining of a headache. There is nothing to indicate that any of the other common symptoms of a serious medical condition were present. Bruce Lee never told Betty Ting, as far as I am aware, that he had problems with his vision, felt like vomiting, and he did not collapse in front of her. This much we can take at face value. However, as Tom Bleecker correctly points out and details in 'Unsettled Matters', the exact truth as to what happened that night was subject to confusion, revision, confabulation and plain outright distortion. This is because of the circumstances relating to where he died and who was present, and not related to the actual cause of his death.

There are some who view Lee's behavior as having become violent and highly unpredictable, and that he was prone to going on what was characterized as 'rampages'. This word, rampage, can be defined as 'A course of violent, frenzied behavior or action.'[29] Some maintain that on the day of his death that this had also occurred and that Raymond Chow was summoned by Linda Lee to calm Bruce down and then it was decided that Lee ended up at Betty's place because he drove him there to seek her help in calming him. It's an odd scenario, someone going on a rampage, and then being taken in a car to the apartment of a single female and then left there. The man involved, Bruce Lee, happens to be one of the greatest martial artists in history. Lee, who is on a 'rampage', the pound for pound strongest and fittest man in the world, compliantly gets into a vehicle, and off they go to Betty's. Then, this man, who is on a 'rampage', is left with this slightly built female, without any regard for her safety, as he is after all on a 'rampage'. What happens, does he assault her? Is the furniture in her apartment trashed? Are the police called by the neighbors due to a disturbance? Are there bottles of booze and empty medication containers strewn throughout the apartment? The answer as we know to all of these questions is no. His 'rampage' consisted of him drinking soft drinks, two cans of soda, one of which was his favorite ginger beer, and there was no evidence of any other drugs in his system except for the marijuana that he had consumed. The other objection I have to the scenario of him being out of control that day is this: *why if he had been out of control, did no one go back to Betty's apartment to make sure she and Lee were alright.* If he had been so out of control that day surely someone, Linda Lee or Raymond Chow or whoever, would have said, "We had better keep a close eye on him and make sure Betty's ok." Especially if, as alleged, the phone was initially unplugged at Betty's apartment. That would have immediately raised concerns. But no, everything was actually under control and Lee was there as intended to spend time with Betty. Those involved got all of their times mixed up at the inquest about who went where at what time because no doubt they had to attempt to avoid scandal and embarrassment. There was nothing suspicious about this at all.

Had he actually been on a rampage it all would have ended a lot differently, the police would have been called and he would have been hauled off, never to be given the Equagesic tablet that led to his death or to be subject to the totally incompetent manner in which he was subject to when he became ill that night by those who should have called an ambulance immediately. Of issue also is whether Lee, when he was at Betty Ting's apartment, was given the Equagesic tablet due to him simply having a normal benign headache, and the Equagesic tablet caused the adverse reaction which led to his death, or did he have the headache symptomatic of the early stages of a cerebral edema and increasing intracranial pressure? It was in all probability a simple headache that Lee had suffered because no other symptoms or signs of cerebral edema were reportedly manifest at that time. There also does not seem to be any evidence that Lee suffered a tonic-clonic (grand mal) seizure at Betty Ting's apartment. Apparently she never reported it. He was a very strong man and on his May 10, 1973 cerebral edema incident the doctors had reported having difficulty restraining him during the tonic-clonic seizure he had at the hospital. For anyone who has seen a tonic-clonic seizure it is dramatic and violent. The individual often injures themselves during the seizure, and objects and furniture around them can be damaged during their thrashing about. Reportedly Betty Ting called Raymond Chow to state that she could not wake Bruce, not that he had had a seizure or other problems. Once it was clear that Lee was ill and those present finally got their act together to call a doctor although an ambulance should have been immediately called. A doctor later arrived and found that Lee was deeply comatose and had a pulse that was 'not acceptable',[30] whatever that means, had no respirations, and his pupils were 'not fully dilated.' In medicine or nursing one can say that a pulse is 'not detectable', or if using a stethoscope 'not audible' or simply compressing an artery 'not palpable'. What does 'not acceptable' mean? If Lee had had a tonic-clonic seizure it should have been clear to the doctor on his examination of Lee, or in the doctor's observation of the immediate environment. The observation about the pupils is interesting. Soon after death, the pupils are slightly dilated, because of the relaxation of the muscles of the iris. Later they are constricted with the onset of *rigor mortis* of the constrictor muscles and evaporation of fluid.[31] Reportedly, when arriving at the Queen Elizabeth Hospital, where he was pronounced dead on arrival, his face was 'swollen like a watermelon.' So I query again, was this due to facial angioedema that occurred due the presumed hypersensitivity reaction that also triggered the cerebral edema? Or was it due to the utterly senseless slapping that he received from those present who instead of wasting time doing that should have been getting him to hospital? The two occurring together, facial edema and cerebral edema, have been documented in the literature and I include here one such case. It centers on an adverse drug reaction to Enalapril, which is used to treat high blood pressure and congestive heart failure. The patient presented to an emergency department with angioedema of the lower face and lip without airway involvement. The Enalapril medication she was taking was discontinued and she was admitted for observation. Several hours later she developed generalized tonic-clonic seizures, which were terminated with intravenous Diazepam and an intravenous loading dose of Phenytoin. Her blood pressure was 170/110 mmHg. There was no evidence of hypertensive retinopathy and her serum creatinine remained normal. Computed tomography (CT) of the brain was performed, which showed extensive cerebral edema involving the white matter. The authors of the paper concluded, 'In conclusion, we have presented a case of diffuse cerebral edema temporally associated with angioedema of the lip and lower face. Alternative causes were excluded and the condition resolved on withdrawal of Enalapril. We propose that Enalapril-induced angioedema of the face and brain was responsible for this presentation.'[32]

The other hypothesis that is reported in the book 'Unsettled Matters' is that Lee died of adrenal crisis. It's an interesting scenario and I refer the reader to Mr. Bleecker's book for his reasoning and further information on this matter. The only problem that I have with this hypothesis is, why, if Bruce Lee suffered from an adrenal crisis which led to his death, or was even causative of the May 10 incident which nearly killed him, that this was not detected and diagnosed at the time by the doctors present? Suppression of the hypothalamic-pituitary axis from chronic exogenous steroid use is the most common cause of secondary adrenal insufficiency.[33] Lee was allegedly using steroids at various points in his life. So Mr. Bleecker's conjecture may not be unreasonable, but there is little direct evidence or testimony by any of those doctors who saw Lee at the time to back the adrenal crisis hypothesis up. There are issues relating to a laboratory test that Lee underwent in Hong Kong which revealed a BUN or Blood Urea Nitrogen level of 92 (normal value 10 – 20 mg/dL) It was queried as to whether this was a typing error and should have read 9.2 mg/dL. We will never know, so this laboratory test is not of any significance because we have doubts as to its validity. Was the reading given actually a typographical error? That's possible. Was it an indication that at that moment in time his kidneys were malfunctioning? That's possible too. But it may not be of particular or enduring significance. Increased BUN levels suggest impaired kidney function. This may be due to acute or chronic kidney disease, damage, or failure. It may also be due to a condition that results in decreased blood flow to the kidneys, such as congestive heart failure, shock, stress, recent heart attack, or severe burns, or to conditions that cause obstruction of urine flow, or to dehydration. BUN concentrations may also be elevated when there is excessive protein breakdown (catabolism).[34] This appears to have been the only occasion in which his BUN was elevated. Following the May 10 incident after getting medically examined in the States reportedly there were no abnormalities with his kidneys or any other pathology detected.

The usual signs of acute adrenal insufficiency include signs and symptoms of patients in acute adrenal crisis such as headache, nausea, abdominal pain, confusion, pale skin, listlessness, dehydration and dizziness.[35] Again, we have to ask, was Lee complaining of, or showing signs of, this specific constellation of signs and symptoms on the night he died? The answer is no. He only complained of a headache. Then some hours later he was dead. This sent the whole process after his death into a tailspin due to the attempts made to avoid embarrassment, avoid scandal and ensure that ever-present priority in Asian culture, the saving of face. Tom Bleecker details all of this in Unsettled Matter and I would encourage readers to review it there. It's just that all of what happened subsequently in regards to that side of things had absolutely no bearing in my view on the medical issues or the causes of Bruce Lee's death.

Lastly, in regards to cerebral edema there is one other issue that we have to clearly review. It hinges on the objection to the purported cause of Lee's death being cerebral edema caused by a hypersensitivity reaction to the Equagesic tablet that he took on the night of his death. There are those who will point out that Lee had an episode of cerebral edema on May 10, 1973 *without having taken Equagesic*. This may be the case, in regards to Equagesic, however he actually had been using one of the two medications that makes up that medication, and that is Meprobamate. The other ingredient in Equagesic is Aspirin. More about this later. I stated at the beginning of this chapter that cerebral edema is symptomatic of various medical conditions and problems, 'cerebral edema, or brain swelling, can be caused by over 300 separate conditions and diseases.' So it is entirely possible to suffer cerebral edema for one reason at one specific point in time and for another entirely different reason at a later date. Let's engage in conjecture and look at other possible conditions causing Lee's episode of cerebral edema on May 10, 1973. Due to his likely dehydrated state the condition of hypernatremia may have been present. This is an electrolyte disturbance that is defined by an elevated sodium level in the blood. Hypernatremia is generally not caused by an excess of sodium, but rather by a relative deficit of free water in the body. For this reason, hypernatremia is often synonymous with the less precise term, dehydration. Water is lost from the body in a variety of ways, including perspiration, imperceptible losses from breathing, and in the feces and urine. If the amount of water ingested consistently falls below the amount of water lost, the serum sodium level will begin to rise, leading to hypernatremia. In some cases, hypernatremia is a result of increased fluid loss via excessive urination. This can be caused by the kidneys failing, which can cause urine output to be dramatically increased. This can also happen because of the use of various diuretics, including loop diuretics and other substances, such as Mannitol.[36] Lee was reportedly using diuretics at various times in his life. We discuss also in this book the issue of Lee having at the time of the May 10 incident an elevated BUN or Blood Urea Nitrogen level. Let's look at this again. There was some doubt, as we have mentioned above, cast on this at the subsequent inquest into Lee's death as it was felt the level recorded at that time, which was 92, may have been a typographical error and actually been 9.2 (a normal value being 10-20 mg/dL) Increased BUN can be caused by dehydration among other things. So it is possible that the test results were actually correct and not a typographical error at all, and fitted in to the clinical picture that presented at exactly that point in time. He had a fever on May 10, his temperature being recorded as 105 degrees. In regards to cerebral edema this can be aggravated by hyperthermia, 'Exposure to a body temperature of near 104°F for a period of only 2 hours increased the edema by 40%. This effect was independent of the hypertensive action of hyperthermia but was intensified when the latter was present. It is suggested that, in human diseases known or thought to be associated with cerebral edema, fever should be vigorously treated.'[37] The point I am making is that there were other factors present that may have led to an episode of cerebral edema on May 10. There are many different causes for this condition, and there are, as we have stated, different types of cerebral edema. Put it this way. Say that I have an episode of fainting on May 10 of one year, and then later that year, on July 20, I have another episode of fainting. Like cerebral edema, there are many underlying reasons for fainting. The two episodes of fainting may be related in terms of cause, but it is entirely possible too that the episodes could occur due to different causes. That is because fainting, as with conditions such as cerebral edema, is symptomatic of various underlying mechanisms, some of which may be multifactorial.

Before further discussing the hypothesis that his episode of cerebral edema on May 10 may have been due to the fever he was running, his likely dehydrated state, and perhaps it being a consequence of attempts at rapid rehydration, let's look at the medication hypersensitivity hypothesis for that date. The fact that we have to remember is that after Bruce consulted a doctor in Hong Kong, in November of 1972, for such issues as acne and excessive sweating he was prescribed, in addition to the other measures taken, a drug called Miltown. This just happens to be Meprobamate and he was therefore prescribed it *prior* to the May 10 incident. We can speculate, and this is all that anyone can do with reference to this particular matter, that perhaps Lee continued using the Miltown, regularly or intermittently, up until the May 10 incident. Is it possible that at that point in time he had ingested a Meprobamate tablet as prescribed and in combination with other factors compromising his physical health at that point, such as dehydration, an elevated temperature and his catabolic nutritional state, that he experienced an acute hypersensitivity reaction to the drug? Following the May 10 incident he refrains from using Miltown again. Then all goes well, until the fateful night of July 20. He complains of a headache and Betty in all innocence gives him a tablet of Equagesic, composed of Meprobamate and Aspirin. The sensitization to the Meprobamate has already occurred, on May 10, and within minutes of him ingesting the tablet on July 20, wham! A serious hypersensitivity reaction occurs, that leads to facial edema and then cerebral edema. Sometimes it is intermittent use of a medication that is the most sensitizing in regards to allergic reactions. 'There is some evidence that sensitization is more likely with higher drug doses and prolonged administration, but clinically this does not appear to be important. *Of greater importance are intermittent courses of moderate drug doses that clearly predispose to sensitization.* '[38] (Author's italics)

The recognized adverse reactions to Meprobamate finally led to the European Medicines Agency recommending suspension throughout the European Union of all medicines containing meprobamate, 'due to serious side effects seen with the medicine.' The Agency's Committee for Medicinal Products for Human Use (CHMP) 'concluded that the benefits of meprobamate do not outweigh its risks.' This was after that agency completed a review of the safety and effectiveness of oral meprobamate-containing medicines, due to serious side effects seen with the medicine.'[39]

And so, in conclusion, the episode of cerebral edema that Lee experienced on May 10 may have been caused by other underlying conditions, such as hyperthermia and dehydration, coupled with his likely compromised nutritional state. Lee was working intensely as usual on May 10 and the air conditioning in the dubbing room had been turned off. The temperature soared. The possible result: 'Heat injury is a common complication of dehydration, which can affect people of all ages and fitness levels. The combination of high temperatures, physical exertion, profuse sweating and insufficient hydration creates conditions that can lead to heat injury. The body uses sweat as a cooling mechanism, in which fluids and electrolytes leave the body and rest on the skins surface. When sweat evaporates, the body cools off. High temperatures and high humidity make this cooling system less effective, meaning that the body continues to sweat (increasing dehydration) but does not cool off adequately. As the body tries to compensate for the climactic conditions, more sweat is expelled, leading to possibly dangerous heat injury. There are various stages of heat injury, each corresponding with increasing levels of dehydration. The first and mildest stage of heat injury is heat cramps. As the body continues to sweat and loses critical fluids and electrolytes, muscles contract slowly, causing painful spasms. The next, and more serious, stage of heat injury is heat exhaustion. This occurs when the body's cooling mechanisms are unable to function properly due to dehydration. As a result, blood vessels and capillaries shut down. Heat exhaustion can cause a variety of symptoms, including headache, nausea, dizziness, loss of coordination and anxiety.'[40] These symptoms match those that Lee experienced on May 10. And further, 'Another severe complication related to dehydration is cerebral edema, or brain swelling. In normal cases of dehydration, patients lose sodium through fluid loss. In cases of hypotonic dehydration, however, too much sodium is lost through sweat. In response to the loss of sodium, the body generates particles to draw extra fluid into the cells. However, if too much extra fluid enters cells, they can swell and rupture. When brain cells are involved, causing cerebral edema, this can be a dangerous complication. Although the brain can adjust to small increases in fluid volume over time, large or rapid increases are not tolerated. Excessive buildup of fluid in the brain can cause brain functioning to be compromised creating a possibly fatal medical emergency.'[41] And further, 'Additional severe complications of dehydration include seizures, kidney failure and coma. Seizures can take place when the body's system of fluids and electrolytes becomes unbalanced, causing electrical impulses in the brain to fire abnormally. Seizures create uncontrollable muscle spasms and can lead to unconsciousness. Without adequate fluids and salts to manage the body's many systems, the kidneys can struggle to evacuate waste products from the body. Therefore, in severe cases of dehydration, kidney failure can occur. Other vital organs and processes can be affected by dehydration as well, resulting in potential comas and fatalities.'[42] And one objective measure of this state is of course, the BUN or Blood Urea Nitrogen level. We could conclude that the laboratory result of Lee's elevated BUN level on May 10 was absolutely correct and caused by his dehydrated state. Once he was rehydrated his BUN levels would have returned to normal, as he actually had no underlying kidney disease. There is one other issue that we must mention and that is the issue of rebound cerebral edema. Rapid rehydration, when the fluid the body has lost is replaced, also has its dangers. 'Neurologic complications can occur in hyponatremic and hypernatremic states. Severe hyponatremia may lead to intractable seizures, whereas rapid correction of chronic hyponatremia (>2 mEq/L/h) has been associated with central pontine myelinolysis. During hypernatremic dehydration, water is osmotically pulled from cells into the extracellular space. To compensate, cells can generate osmotically active particles (idiogenic osmoles) that pull water back into the cell and maintain cellular fluid volume. *During rapid rehydration of hypernatremia, the increased osmotic activity of these cells can result in a large influx of water, causing cellular swelling and rupture; cerebral edema is the most devastating consequence'.*[43] (Author's italics) Slow rehydration over 48 hours generally minimizes this risk. So physiologically, dehydration, despite the name, does not simply mean loss of water, as water and solutes (mainly sodium) are usually lost in roughly equal quantities to how they exist in blood plasma. In hypotonic dehydration, intravascular water shifts to the extravascular space, exaggerating intravascular volume depletion for a given amount of total body water loss. Neurological complications can occur in hypotonic and hypertonic states. The former can lead to seizures, while the latter can lead to osmotic cerebral edema upon rapid rehydration. Again, was it possible that on arriving at hospital on May 10 the doctors recognized Lee's dehydrated state. Receiving rapid rehydration through parenteral (I/V) fluids he suffered an osmotic cerebral edema, a 'rebound' cerebral edema. Again I am at a disadvantage because I do not have access to any of the detailed medical notes surrounding these events. I am engaging in speculation, but that is all I and the other authors who write about Bruce Lee and these specific medical matters can do. I have included in this book detailed facts and information in regards to cerebral edema for a reason. Despite the fact that such matters may seem dry and difficult to understand, for all of us at times, it is important to detail it all so that those reading this book can come to a conclusion regarding these processes themselves. We have to have an exact and precise explanation of what happened, and our thinking and reasoning has to be founded on good science, good research and sound reasoning. This is why I have methodically included detailed information in regards to these clinical syndromes so that readers can study these and have an understanding of what exactly happens in such situations and why. I welcome any rational criticism of my assertions and statements.

The occurrence of the episode of cerebral edema on May 10 does not of itself present any difficulties clinically in accepting that another episode of cerebral edema occurred on July 20. We can hypothesize that the episode of cerebral edema that occurred on May 10 was due to the circumstances and his dehydrated and hyperthermic state, or perhaps the sub-type called osmotic cerebral edema due to rapid rehydration. The literature supports such a contention, 'We speculate that these metabolic perturbations may relate to the development of cerebral edema and seizures or coma following rapid rehydration of humans with chronic hypernatremic dehydration.'[44] The episode on July 20 may have been a cerebral edema of a different sub-type. Whether these episodes were due to entirely different underlying causative factors, or whether they were due to Lee using Meprobamate on May 10, and him experiencing an initial hypersensitivity reaction at that time we do not know. Then, in regards to the Meprobamate hypothesis, he discontinued his use of the drug until it was given to him again on the night of his death and he then suffered a further catastrophic hypersenstitivity reaction. This we can speculate, however we will never know with complete certainty. However these events happened and they may easily have happened for either reason.

Chapter 3
Cryptorchidism – the role that this played in the life of Bruce Lee.

In 'Unsettled Matters', the author, Tom Bleecker, very early in the book gets into the fact that Bruce Lee had a physical abnormality. The disorder is considered significant from the point of view of there being a high risk of testicular cancer in males with this condition. Mr. Bleecker points out that Bruce himself was obviously sensitive about it and mentioned it often to others. The real significance in my view of the fact that Bruce Lee suffered from an undescended testicle, or cryptorchidism, is that it had, in his early years, an enormous *psychological* impact on him. According to the many life histories of Lee he was constantly getting into trouble for fighting, when at school and when outside of school. In contrast to other boys it is likely that this behavior may have been driven by a feeling of inadequacy, and a continual demand to prove himself as a man. This may at least in part explain what was often referred to as his insufferable cockiness. It was part of his character and it hid very deep feelings of inadequacy. In fact, as we know, there was nothing inadequate about Bruce Lee. He was referred to by the martial artist Ed Parker as 'a one in a two billion' man.[1] Ed Parker was right. He was an extraordinarily talented and gifted man. It's just that psychologically he felt compelled and driven always to prove it. This can have both adaptive and maladaptive effects. In the end it became maladaptive and his obsessive drive for perfection would possibly be a factor in his death, in the way that he neglected a well-balanced diet and attempted, at least at times, to survive on a fluid diet. There were essentially two critical psychological complexes that drove Bruce Lee's behavior. One was this deep feeling of inferiority that stemmed in part from his feeling that somehow he didn't match up to other men, that something was 'missing.' With such a condition you would come in for some bad teasing in the changing rooms at school for instance. And that may explain some of his fights at school. And also, the question was bound to go through one's head, *just how will women view me?* For anyone with any admiration for Bruce Lee it didn't matter at all, but *for Bruce* it obviously mattered and he was deeply insecure about this. And it seemed to play out in his early romantic relationships. The other key to his personality and the way that he was driven was his cultural background. Mr. Bleecker in his book rightly and appropriately puts things in context by discussing the political and cultural issues of Hong Kong that were present during Bruce's formative years. Hong Kong was administered then by the British, who were perceived, by themselves at least, to be 'superior' and were paternalistic in so many ways. 'Orientals' were subjected to mistreatment or just plain benign neglect, as Bruce incorporated into his movies. But of course the real issue was that for Lee the drive to overcome this would prevail. And so he was driven, as an Asian, an 'Oriental', to prove himself superior or at least equal. To his father, his family and the world. And reinforcing this was the Chinese cultural trait of rising to the top, becoming rich and a success. And if and when that happened, *flaunting that success at every opportunity.* Bruce Lee was every inch an Asian and he embodied all of the cultural, emotional and familial traits and complexes that that implied.

In terms of the formal research on the condition of cryptorchidism there is a small body of research on the psychology of this condition that attempts to determine whether it can cause lasting psychological problems. This research consists of only a few case reports and small studies. This research also suffers from serious methodological problems: major variables are completely uncontrolled, such as the small physical stature of many cryptorchid boys, and the psychological effects of corrective surgery. Despite these research limitations, a few important results do emerge. Several psychoanalytic case studies indicated that cryptorchidism by itself would not produce psychological disorder; but when combined with a common distinctive family pathology it did produce typical symptoms.[2] The most striking dynamic in this small sample were the parents' contradictory attitude to the boy's problem: a worried preoccupation with the genitals—constant poking, checking, and medical exams—conjoined with blatant denial that anything was wrong. The mothers were controlling and discouraged boyish aggression; the fathers were remote and dissatisfied with their son whose physical defect was equated to their own personal failures. In a research study similar family patterns were observed. The mothers of these cryptorchid boys tended to be possessive and regarded them as inadequate. The fathers were uninvolved and disparaging, and worried that the boy was a sissy or a freak.[3] Similar conclusions were reached in the research study by Cytryn, et al. in 1967. The boys' figure drawings showed poor sexual differentiation, were often missing limbs, or portrayed males as inferior to females.' Over half the boys '....seemed to see themselves as masculine but maimed or incomplete in some essential way.'[4] A related finding is that a thrust of masculine development is evidenced in cryptorchid boys upon repair of the undescended testicles. In each of Blos' cases corrective surgery produced a euphoric upsurge of male sexuality followed by major advances in assertiveness, learning, and socializing. A similar outcome was reported with a 9-year-old Puerto Rican boy. In another case study, a cryptorchid man in his twenties had successful surgery: his operation unleashed a wave of adolescent male strivings, as though the testicles represented, '....the masculine resources he always had within himself if he only had the courage to express and assert them.'[5]

As an aside, many other men have suffered from this condition, one of those commonly being included in this group being Adolf Hitler. Coincidentally, while completing this book, I was reading at the time the German language version of the definitive biography of

Hitler by John Toland. During the time he was in his bunker in Berlin, prior to the collapse of the Third Reich, Hitler had various ailments, most of which were psychosomatic. His doctors pumped him full of various drugs, with stimulants to counter his growing lethargy and despondency interspersed with hypnotics to help him sleep. A Dr.Giesing had the opportunity at one point to physically examine Hitler and the book continues (Author's translation from German) 'He was convinced by this opportunity, that the rumors over the abnormally developed sexual organs of Hitler were contrary to the facts; in this regard Hitler was bodily intact and completely normal.' The book states that at least two other physicians who had had the opportunity to examine Hitler arrived at the same conclusion. One of these was another of his personal physicians, a Dr. Morell, who also reported that Hitler's sexual organs were completely normal.[6] The rumors of Hitler suffering cryptorchidism have been around for a very long time, but they are not true.

So Bruce Lee did suffer from this disorder called cryptorchidism. One of his testicles had failed to descend into the scrotum prior to birth. It was one of the medical issues that led to him being rejected by the military draft:

'At the end of the summer and upon his return to Seattle, he [Bruce Lee] found his draft papers waiting. He went to the Induction center but was amazed to find himself rejected by the U.S. Army, classified as "4F" due to an apparently undescended testicle, poor eyesight, and a sinus disorder. Bruce was somewhat bemused to be the fittest man the Army ever rejected; however, he did eventually don a uniform as a member of the campus ROTC squad.'[7]

We have looked at the psychological aspects of this condition; now let's look at cryptorchidism from a medical and physiological point of view. I want to do this because in Unsettled Matters Mr. Bleecker states that when Bruce in his adult years attended a physician, that this physician, '....was unaware that Bruce's adrenal glands, which are responsible for the production of cortisone, as well as the anabolic steroid testosterone, had been producing at a substandard level since Bruce was in his mother's womb.'[8] Let's just read that again, '....Bruce's adrenal glands, which are responsible for the production of cortisone, as well as the anabolic steroid testosterone, had been producing at a substandard level since Bruce was in his mother's womb.' How do we know this? Stated also in Unsettled Matters, when reporting his cryptorchidism, is that it is a condition '....wherein one or both of the newborn's testicles have failed to descend prior to birth. The condition is often routinely corrected by the administration of testosterone (an anabolic steroid), which the infant and its mother failed to produce through their adrenal glands in the infant's eighth fetal month.' And how do we know this also? The author of 'Unsettled Matters' had access to the medical records of Bruce Lee. Was there in those records a blood test result that showed unequivocally that Lee was deficient in testosterone and cortisone? If there was there can be no argument. That would be conclusive objective evidence that this was the case. It may be though that Mr. Bleecker is basing the above statement on the one of the hypotheses that relates to the presumed causes of cryptorchidism. I hope this can be clarified in due course. We are discussing the biography of a significant figure in modern-day history and culture. We must pay Bruce Lee due respect in this regard and we must get the facts of his life and his medical history documented with absolute precision. So it becomes here a point of discussion and debate as to what the precise cause of his cryptorchidism was and whether this condition led to him producing hormones subsequently at a 'substandard level.'

Let's look at the literature regarding firstly testosterone.

'Testosterone is a steroid hormone from the androgen group and is found in mammals, reptiles, birds, and other vertebrates. In mammals, testosterone is primarily secreted in the testicles of males and the ovaries of females, although small amounts are also secreted by the adrenal glands. It is the principal male sex hormone and an anabolic steroid.'[9] Testosterone is necessary for normal sperm development. It activates genes in *Sertoli cells*, which promote differentiation of spermatogonia. We know that Bruce Lee was fertile and virile. There was no evidence in this regard that he had 'substandard' levels of testosterone. For those who may be interested testosterone levels can fluctuate throughout life in normal men. Fatherhood also decreases testosterone levels in men, suggesting that the resulting emotional and behavioral changes promote paternal care.[10] Also, and this is interesting, although it's something that we know without the scientific research, 'Men producing more testosterone are also more likely to engage in extramarital sex' [11]

There is another separate condition that we have to mention and that is hypogonadism, and this condition is one that, as far as I am aware, was definitely not reported in Bruce Lee's medical records. Again, with regards to this, researchers such as myself are at a distinct disadvantage because we have not been able to access these. As has been stated, one function of the testes is to secrete the hormone testosterone. This hormone plays an important role in the development and maintenance of many male physical characteristics. These include muscle mass and strength, fat distribution, bone mass, sperm production, and sex drive. Cryptorchidism and hypogonadism are two separate conditions. Hypogonadism in men is a disorder that occurs when the testicles (gonads) do not produce enough testosterone. Primary hypogonadism occurs when there is a problem or abnormality in the testicles themselves. Secondary hypogonadism occurs when there is a problem with the pituitary gland in the brain, which sends chemical messages to the testicles to produce testosterone. Hypogonadism can occur during fetal development, at puberty, or in adult men. When it occurs in adult men it may cause the following problems:

Erectile dysfunction (the inability to achieve or maintain an erection)
Infertility
Decreased sex drive
Decrease in beard and growth of body hair
Decrease in size or firmness of the testicles
Decrease in muscle mass and increase in body fat

Enlarged male breast tissue

Mental and emotional symptoms similar to those of menopause in women (hot flashes, mood swings, irritability, depression and fatigue)[12]

Let's now look at some research from the medical literature in regards to cryptorchidism and what causes it:

One study in regards to the condition found that 'Cryptorchidism is one of the few well-described risk factors for testicular cancer. It has been suggested that both conditions are related to increased *in utero* estrogen exposure. The evidence supporting the 'estrogen hypothesis' has been inconsistent, however. Testicular descent is a two-phase process that takes place in both first and third trimesters, hormone levels in the third trimester may be as important in understanding the etiology of cryptorchidism as hormone levels earlier in pregnancy. An alternative hypothesis suggests that higher *in utero* androgen exposure may protect against the development of cryptorchidism and testicular cancer. The results [of this study] found no significant differences in the levels of testosterone (total, free, bioavailable), α-fetoprotein, sex hormone–binding globulin, or in the ratios of estrogens to androgens. Total estradiol, however, was significantly lower in the cases versus the controls ($P = 0.03$) among all mothers and, separately, among White mothers ($P = 0.05$). Similarly, estriol was significantly lower among all cases ($P = 0.05$) and among White cases ($P = 0.05$). These results do not support either the estrogen or the androgen hypothesis. Rather, lower estrogens in case mothers may indicate that a placental defect increases the risk of cryptorchidism and, possibly, testicular cancer. Higher testosterone levels among Black mothers in comparison with White mothers have also been reported by Troisi et al. and Zhang et al. Indeed, the current study also found that Black mothers had higher testosterone levels than White mothers. Similarly, in comparing Chinese mothers to White mothers, Lipworth et al. reported higher testosterone levels in Chinese mothers. Like Black men, Chinese men have much lower incidence of testicular cancer than do White men'[13]

Beyond the hormonal hypothesis in regards to the etiology of cryptorchidism is this, 'Evidence supporting genetic causes of cryptorchidism is abundant. In some cases of unilateral cryptorchidism, the contralateral, normally descended testis may be altered, too, and testicular cancer may originate from the contralateral, not retained testis. These findings suggest that cryptorchidism might be considered a sign of an underlying congenital alteration of the testes.'[14]

Further,

'....long term sequalae of cryptorchidism may occur, including infertility, subfertility, and malignancy. (Bruce Lee was apparently *not* infertile or subfertile – author) Three percent of full-term boys and up to 45% of preterm males have cryptorchidism. This percentage falls to 1% by 3 months of age. There are two peaks for detection of undescended testes: at birth, and at 5 to 7 years of age. The latter group probably represents those patients with low undescended testes ('ascending testes') that become apparent with linear body growth. Bilateral undescended testes occur in 10% of patients with cryptorchidism. Unilateral anorchia is found in 5% of patients. An undescended testis is one that does not remain at the bottom of the scrotum after the *cremaster muscle* has been fatigued by overstretching. This is commonly confused with a retractile testis, one that may not always lie in the scrotum, but that will stay in the bottom of the scrotum after overstretching the *cremaster muscle*. Advanced maternal age, maternal obesity, diabetes, or consumption of cola-containing drinks during pregnancy, family history of cryptorchidism, prematurity, breech presentation, and low birth weight/small for gestational age have all been suggested as possible risk factors for cryptorchidism. Normal testicular descent occurs during the seventh month of gestation. The majority of testes that will descend spontaneously do so by 3 months of age, possibly due to the normal gonadotropin surge (luteinizing hormone [LH] and follicle-stimulating hormone [FSH]) that occurs around 60-90 days of life and is responsible for maturation of the germ cells. Gendrel et al. and Job et al. reported blunting of this surge in boys with cryptorchidism that remain undescended in the first year of life. They demonstrated a significant difference in the polynomial regression curves comparing the testosterone levels of persistently cryptorchid testes with those having delayed spontaneous descent. Many boys with cryptorchidism have lower morning urinary LH and a decreased LH/FSH response to gonadotropin-releasing hormone, corresponding to the abnormal germ cell development in both the undescended and contralateral descended testis. Without this surge, Leydig cells do not proliferate, testosterone does not increase, germ cells do not mature, and infertility may result. This suggests that a mild endocrinopathy is responsible, and cryptorchidism may be a variant of hypogonadotropic hypogonadism. *However, the positive predictive value that bilateral cryptorchidism will have abnormally low testosterone levels is only about 23%* (Author's italics).   Subsequent studies have been inconsistent, with reduced or normal serum testosterone levels and normal or relatively increased relative LH levels, when comparing cryptorchid patients with controls. Of all boys with undescended testes, 4% of their fathers and 6-10% of their brothers also have undescended testes. *Though a genetic susceptibility for cryptorchidism is suggested, the etiology is likely polygenic and multifactorial.* (Author's italics) Schnack et al. reviewed over 1 million male births and demonstrated that risk ratios for cryptorchidism were 10.1 in twins, 3.5 in brothers, and 2.3 in offspring of fathers who had an undescended testis. Previous data on this subject suggested 5-fold increased risk in offspring of affected fathers and 7- to 10-fold increased risk in those with an affected brother as compared to patients with no family history of the disorder.[15]

And another study, which states,' Complete testicular descent is a sign of, and a prerequisite for, normal testicular function in adult life. The process of testis descent is dependent on gubernacular growth and reorganization, which is regulated by the Leydig cell hormones insulin-like peptide 3 (INSL3) and testosterone. Investigation of the role of INSL3 and its receptor, relaxin-family peptide receptor 2 (RXFP2), has contributed substantially to our understanding of the hormonal control of testicular descent.

Cryptorchidism is a common congenital malformation, which is seen in 2–9% of newborn boys, and confers an increased risk of infertility and testicular cancer in adulthood. *Although some cases of isolated cryptorchidism in humans can be ascribed to known genetic defects, such as mutations in INSL3 or RXFP2, the cause of cryptorchidism remains unknown in most patients'.*(Author's italics)

The point I am making is that if the cause remains unknown we cannot state that it is explained by a hormonal hypothesis, and therefore we cannot infer or deduce from that that Bruce Lee was 'deficient' in any way in regards to his physiology. Further, 'Several animal and human studies are currently underway to test the hypothesis that in utero factors, including environmental and maternal lifestyle factors, may be involved in the etiology of cryptorchidism. Overall, the etiology of isolated cryptorchidism seems to be complex and multifactorial, involving both genetic and nongenetic components.'[16]

And so, are we here making an assumption that Bruce Lee had, in Mr. Bleecker's words, 'substandard levels', of cortisone and the anabolic steroid testosterone? The causes for cryptorchidism are actually poorly understood and may actually be related to mechanical rather than hormonal causes. Let's go head to head with Mr. Bleecker on this. We can discuss this rationally and have an informed debate. Mr. Bleecker may well be right, but I simply want to review the reasoning and the statement that was made. In Unsettled Matters he states that in addition to lifting weights Lee was using high energy protein drinks to build added muscle but that these were not working. The reason, according to Mr. Bleecker, '….was due to Bruce's low level of testosterone that resulted from his childhood cryptorchidism (caused by his low performing adrenal glands) and subsequent removal of one of his testicles.'

Let's have a further look at what the medical literature states.

A study by De Muinck Keizer-Schrama et al. found no differences in basal or stimulated LH, FSH, and testosterone levels or stimulated dihydrotestosterone or testosterone precursor levels among 29 persistently cryptorchid boys, 18 with spontaneous descent until 6 months, and 144 controls followed longitudinally from 3–12 months. They studied pituitary-gonadal function during the first year of life in 48 boys born with 56 undescended testes in order to test the hypotheses that functional insufficiency of the hypothalamo-pituitary-gonadal axis and disorders of testosterone (T) biosynthesis occur in such boys. Cryptorchidism persisted for longer than 1 yr in 29 boys (30 testes; group I), whereas spontaneous descent occurred in 19 boys (20 testes; group II), in 6 after the sixth month. A control group (group III) included 160 boys. Basal and peak LHRH-stimulated serum LH and FSH and basal serum T values were determined at 3, 6, and 12 months. Serum T, dihydrotestosterone (DHT), progesterone (P), 17-hydroxypregnenolone, 17-hydroxyprogesterone, dehydroepiandrosterone sulfate, and androstenedione before and after hCG administration were determined at age 1 yr. Comparing the 3 groups, cross-sectional evaluation revealed no significant differences in basal or peak LHRH-stimulated serum LH and FSH levels, except that basal serum LH levels were slightly higher in group II than in group III. Comparing groups I and II, longitudinal evaluation revealed similar basal and peak LHRH stimulated serum LH and FSH values, with comparable changes with time. Basal serum T, DHT, and T precursor levels were similar in all three groups, with similar rises of T and DHT and variable minimal increases in androstenedione and dehydroepiandrosterone sulfate after hCG stimulation. *We conclude that during the first year of life, boys with cryptorchidism have no functional insufficiency of the hypothalamo-pituitary-gonadal axis or disorders in T[testosterone] biosynthesis.*[17] (Author's italics)

According to Unsettled Matters Lee went on to use steroids and also diuretics. Mr. Bleecker correctly points out the dangers of these two substances, and correctly reports in his book the long term adverse effects of steroid use. The use of diuretics is, as Unsettled Matters points out, fraught with danger. The reason for this is that due to the rapid loss of fluid from the body all sorts of problems can occur. And it is not only the fluid loss, but the loss of essential electrolytes such as potassium that are essential for correct heart functioning. There have been a number of deaths from individuals inappropriately using diuretics and suffering fatal cardiac arrhythmias. This is particularly so for athletes such as Lee who engaged in high intensity exercise regimens and incurred the further natural loss of potassium through sweating.[18]

Bruce Lee may well have used steroids. I really cannot contradict Mr. Bleecker on this. But I do object to some of the evidence that he puts forward in his support of this claim. He writes that there is a specific psychological profile that users of anabolic steroids manifest, and that this includes '….an increase in self-esteem to the point sometimes of severe narcissism and self-obsession. Further that there is an increased drive to exercise intensely, and a decrease in their ability to accept poor performance or failure, and a marked decrease of tolerance for others.' This supposedly aptly describes Lee in the early 1970's. Well, it does, but from what we know it aptly describes Lee *all of his life*. And self-obsession and narcissism is extremely common in 95% of those with the postal address Hollywood, whether on steroids or not. Lee had *not* apparently been exercising in the last weeks of his life, and reportedly wanted to focus on acting and producing movies. So his drive to exercise intensely appeared to have waned, not increased. He could never accept poor performance or failure, but that was part of his personality. And he had a hot temper and would always challenge those whom he felt were fakes. This cluster of personality traits does not in my view of itself lend any specific weight to the assertion by others that he was using steroids. Although, as I have said, there may well be other incontrovertible evidence that he was.

This all relates to Lee's behavior just prior to his death and how his alleged use of steroids may have influenced his behavior and made him become erratic and violent. We go on later to read in 'Unsettled Matters' about an altercation that Lee had with a man by the name of Lo Wei. Mr. Bleecker writes that, 'With synthetic male sex hormone now soaring in his bloodstream, 'Bruce's 'roid raging had returned.' Well, this could be explained, rather than resorting to such claims as 'roid raging', with the fact that they hated each other's guts and stirred each other up no end? There was a big argument with Lo Wei one day. Lee abused him, Lo Wei and his wife gave back as good as they got. Wei's wife was obviously not terrified of Lee and apparently gave him a good verbal clobbering in defense of her husband. It was all a storm in a Chinese tea pot. Lee didn't assault Wei and no one was murdered. The police were called who recognized it was just two people having a plain old set to. Did the police charge Lee, or attempt with reinforcements to haul him off for a psychiatric examination with the thought, *this guy Lee is really off his head and out of control, we had better take him to a doctor?* No. They asked him to sign a form agreeing to desist from threatening Lo Wei. Which he did, and Lo Wei, as far as I am aware never suffered so much as a scratch, although Lee reportedly threatened him with a knife. I don't condone such behavior, but it may not have been an episode of 'roid raging' at all. The alternative hypothesis is that it was simply the result of a long-simmering feud and Lee expressing all of his frustrations and stresses regarding his life at the time. All over the world, in offices, in homes, and everywhere else people go head to head and have big, loud arguments where they scream and shout and threaten to murder each other, but never actually do. In the vast majority of these cases it has nothing whatsoever to do with 'roid raging' and I don't believe in this case it necessarily had anything to do with 'roid raging' either.

There is other evidence that paints a picture that is completely contrary to the insinuation that Bruce Lee was out of control and violent. The story, and it is reportedly a true one, was that Bruce Lee was challenged on a movie set while he was filming in Hong Kong. It was related by another martial artist. The young man was sitting on a wall and abused Lee and told him (reportedly in Chinese – the witness who is not Chinese found out later what the man had said) that Lee was a phony and was not a good martial artist. Apparently they were between takes and Lee challenged the man to come down and beat him up! The witness reports that Lee's challenger was no slouch and really attempted to beat Lee up. Lee, we are told, methodically took the challenger apart, trapped his hands, slammed him against the wall a number of times and in the end lectured the man on such issues as his stance, which Lee told him was too wide! The challenger, who as a consequence of Lee's superb skill and self-control was defeated without sustaining injury, then shook Lee's hand and said "You really are a master of the martial arts." So, Lee was completely in control, even in this situation, and the witness to this event stated, "Bruce not only punched the kid out, but he didn't hurt him doing it. I was very impressed."[19] Lee in complete and total self control, mastering a difficult situation and managing to dominate the challenger completely without injury to the opponent. So much for 'roid raging' and out of control behavior. There may always be an equally valid contrary view or a different way of explaining things. Out of respect for Bruce Lee we should give both sides of any story and not characterize an event as 'roid raging' when perhaps it wasn't. As for the argument with Lo Wei there was no shortage of this kind of drama going on every day in Hong Kong. It was a high expressed-emotion place. Reportedly Lee wanted to end the relationship with Ting, but she had some sort of emotional breakdown and was admitted to hospital. It was all high drama in true life. And apparently Lee had taken to drinking. Who wouldn't under such circumstances! But the night of his death Bruce Lee was as sober as a judge, and all was deathly quiet at Betty Ting's apartment. No fights, no rampage, no police being called, no ambulances until the very end. Bruce Lee was not out of control, although there have been attempts made to portray him in this way.

The only way we would know with any certainty if 'Bruce's adrenal glands, which are responsible for the production of cortisone, as well as the anabolic steroid testosterone, had been producing at a 'substandard level' since Bruce was in his mother's womb would be via a blood test and laboratory measurement of these hormone levels in his bloodstream. Did this happen? I for one would really like to know.

Therefore, by reviewing the medical literature, we can see that this assumption can be challenged because the hormonal hypothesis is only one possible explanation for a condition that is not fully understood. Bruce Lee's hormone levels may have been perfectly normal. His undescended testicle, and the etiology of his cryptorchidism, may have been explained by other factors and processes. I can only ask that we be given, or clear reference is made, to the hard evidence substantiating the claim that Bruce Lee, at the time he consulted this physician, had 'substandard' levels of the specific hormones that is mentioned. It is a crucial link in the chain of assumption and inference made, leading all the way up to the actual death of Lee.

In any case, and no doubt to reduce the risk of testicular cancer that this condition presents with, in March of 1969, in St. John's hospital in Santa Monica, surgeons removed the undescended testicle.[20] I am not an endocrinologist but it doesn't sound to me as if Bruce Lee was deficient in testosterone. He was always getting into fights, and was fully virile and fertile. So it sounds to me that he was as normal and as testosterone driven as any other hot-blooded male!

Chapter 4
Back injury? What back injury?

There is another issue regarding his medical history that we should examine and discuss. It relates to his well-known back injury. In July of 2012, as part of my research, I accessed the biography section for Bruce Lee on the website of The Bruce Lee Foundation. This is what I read:

'Bruce was devoted to physical culture and trained devotedly. In addition to actual sparring with his students, he believed in strenuous aerobic workouts and weight training. His abdominal and forearm workouts were particularly intense. There was rarely a time when Bruce was doing nothing—in fact, he was often seen reading a book, doing forearm curls and watching a boxing film at the same time. He also paid strict attention to his food consumption and took vitamins and Chinese herbs at times. It was actually his zealousness that led to an injury that was to become a chronic source of pain for the rest of his life. On a day in 1970, without warming up, something he always did, Bruce picked up a 125-pound barbell and did a "good morning" exercise. That consists of resting the barbell on one's shoulders and bending straight over at the waist. After much pain and many tests, it was determined that he had sustained an injury to the fourth sacral nerve. He was ordered to complete bed rest and told that undoubtedly he would never do gung fu again. For the next six months, Bruce stayed in bed. It was an extremely frustrating, depressing and painful time, and a time to redefine goals. It was also during this time that he did a great deal of the writing that has been preserved. After several months, Bruce instituted his own recovery program and began walking, gingerly at first, and gradually built up his strength. He was determined that he would do his beloved gung fu again. As can be seen by his later films, he did recover full use of his body, but he constantly had to take measures like icing, massage and rest to take care of his back.'[1]

The Bruce Lee Foundation records its vision as, '.... to educate and inspire young people using Bruce Lee's life example and philosophies. We will do this through a thriving scholarship program and educational outreach programs that inspire and motivate as well as through our website, books, articles, newsletters and videos. Not only are we helping individuals, who show a dedication to educating and expressing themselves and who may need assistance in their process to reach their goals of higher education, but we are helping early on to build that dedication to education and self-expression through our educational programs.' Excellent and very worthy goals and I fully agree with them.

Was Bruce Lee in bed for six months? Does the foundation assist with authors and researchers such as myself who seek to find out about these matters? I did email the Foundation twice to attempt to open a discussion regarding this, at two email addresses given on the website, but I did not receive a reply and the emails did not bounce back as undelivered. Perhaps the Foundation is inundated with emails and they obviously are doing some very good work in regards to scholarships and promoting the martial arts.

Let me say at this point that the death of Bruce Lee and the death of Brandon Lee were tragedies, and I cannot imagine how Linda Lee must have felt to have lost a husband and a son so suddenly, and so needlessly. For Shannon Lee it was her father and then later her brother that were taken from her. I have the greatest admiration for them and I must try to treat the matters that I deal with as sensitively as I can. With all significant men of history and of popular culture, which Lee was, we owe it to their memory to provide a completely accurate and properly detailed and referenced biography. And as I have stated I initiated contact with the Foundation to attempt to discuss and clarify these issues with no response. With regards to the back injury pretty much everywhere the same old story is churned out and repeated:

'Lee's intense training though eventually led to injury. He severely injured his fourth sacral nerve while weightlifting without warming up in 1970. Afterwards, he was basically bedridden for six months. While he was told he would never practice martial arts again, the Bruce Lee Foundation biography states that he, "instituted his own recovery program... gradually built up his strength... [and] as can be seen by his later films, he did recover full use of his body." [2]

And,

'Bruce Lee injured his back causing damage to his sacral nerve in 1970. The injury was due to overtraining and lifting too heavy during "Good Mornings", a weight training exercise, not during a fight as many people believe. Although doctors told him he would not be able to continue his lifestyle in the martial arts, through determination he fully recovered and went on to star in four and a half films made between 1971 and 1973.'[3] Excuse me, but which specific doctors stated this?

In the fictional movie, Dragon, we get the following account posted on Wikipedia:

'Bruce wins the fight but Sun attacks Bruce from behind after the fight is over, resulting in a serious back injury. While immobilized and recovering Bruce and Linda quarrel of why he did not tell her about this duel, but she furiously rejects his despairing assumption that she will abandon him because of this injury. To give his recovery time purpose, Linda coaxes him to examine the weaknesses of his combat technique, which leads to him developing the fight philosophy of Jeet Kune Do while she helps him write 'The Tao of Jeet Kune Do.' During this period Linda gives birth to their first child, Brandon, and he is the key for the couple to reconcile with Linda's mother. Later at Ed Parker's martial arts tournament, Bruce has a new face-off with Johnny Sun, this time in a 60-second demonstration of his new fighting style. Johnny Sun appears to have the upper hand in the first half of the match but then Bruce recovers and ends up kicking Sun over the top rope.' Further, 'Wong Jack Man did not kick Bruce Lee in the back while Lee was walking away from the fight. The fight did in fact take place, but it was at Lee's own school, not at the strange temple seen in the film. Bruce won successfully, but his fighting style was very limited at the time. This fight was the reason that Lee would develop his own style, Jeet Kune do. Lee's back injury is based on his months long hospitalization due to pinching his sacral nerve while doing weighted good-mornings. In the movie Wong was named Johnny Sun.'[4] And herein is part of the problem. The truth has been mixed in with the fiction, and it becomes increasingly difficult for fans such as myself to know what in reality happened.

And again on Wikipedia:

'The good-morning is a weight-lifting exercise. The movement resembles bowing to greet someone 'good morning'. The *erector spinae* muscles of the lower back work isometrically to keep the spine in an extended position while the hamstrings and *gluteus maximus* work isotonically to perform hip extension. Other muscles are involved in stabilizing weight on the back and maintaining balance. The good-morning is a controversial exercise as some will claim that it leads to lower back injuries. Famously, Bruce Lee seriously injured himself while performing the exercise after an inadequate warm-up and overconfidently selecting his working weight. On the other hand, the good-morning can also strengthen the lower back and prevent injury when properly applied.'[5]

And the following:

'On August 13, 1970, Lee was performing Good Mornings with 60kg / 135lb - his bodyweight at the time - and was completing his first set of 8 (he usually did 2 sets) without sufficiently warming up when he heard a loud popping sound, and dropped the weighted bar. For several days he tried heat treatments and massage, until the steadily increasing pain forced him to seek medical advice. He had severely damaged a 4th sacral nerve, and it was unlikely that he would ever be able to kick again; in fact walking unaided was in doubt. He was forced to rest, and for the next six months he spent most of his time either lying or sitting up reading from his extensive library. During this time he also designed a bed which would afford him greater comfort in his injured state. Eventually he resumed teaching and training, not because he was fully healed, but simply as he felt he had given himself enough time and was unable to refrain from his active life any longer. He would suffer chronic back pain for the remainder of his life, and began taking marijuana as this helped numb the pain.'[6]

As a Bruce Lee fan, and as someone interested in the historical truth of the biography of this extraordinary man, when reading all this I had the thought, *would someone tell me what the hell is going on? Did he have a back injury or didn't he?* How exactly did it happen and to what degree was he incapacitated by it both short term and long term? In pretty much everyone's mind it is a matter that is mentioned over and over again, about this back injury that seems to have never actually been present. Even those who knew Bruce extremely well, including martial artists who were close to him, have commented on this:

"In 1968 [Bruce Lee] was heavily into weights" And then he notes that concerning his back injury, 'There were so many rumors." "He was lifting weights one night with a student. He then suffered a "ruptured disk." Lee spent "three weeks in hospital" and was "in constant back pain." They, "didn't know if he would walk again or not", however, "he was up within a month." He goes on to then state how Lee's death occurred. Lee reportedly took medication for a headache which, "….reacted with the medication for his back." It is correctly reported that Lee died of cerebral edema, a swelling of the brain, and that this, "ruptured blood cells in his brain."[7]

So we have two issues here that we mention above and that are also mentioned by the Bruce Lee Foundation website, and by one of his martial arts contemporaries. These are the back injury and the cause of Lee's death, the cerebral edema. Let's go back at this stage to briefly review our knowledge concerning cerebral edema because it is important that we fully understand this condition:

Cerebral edema is when the intra- or extracellular layers of the brain become flooded with excessive water. Cerebral edema is also known as brain edema. There are four basic distinct types of cerebral edema. Brain edema is a pathophysiological response that occurs at the cellular level and manifests in an excessive concentration of water in brain. Damaged cells swell and leaking or ruptured blood vessels block absorption pathways within the brain, causing excess fluids to breach the blood-brain barrier. A brain injury cascade ensues as a consequence of cell and vessel damage. Calcium and sodium entry channels open as a result of the influx of glutamate around the cerebral membranes. Calcium ions are exchanged for sodium ions within the cell. The buildup of sodium within the cells creates an osmotic gradient which then results in a release of water, which causes the cells to increase in volume. Cerebral edema causes brain malfunction but does not necessarily cause permanent brain damage. The pathophysiological feedback loop finally ends as a result of the hypoxia which triggers the opposite response reversing the swelling. All episodes of cerebral edema are temporary. However, the damage this pathophysiological response can inflict can be life threatening if effecting important nerve centers in the brain.

A brain edema that is caused by ruptured cerebral vessels is known as vasogenic edema. Vasogenic edema occurs as a result of the breakdown of the tissue in the blood-brain barrier. The blood brain barrier prevents foreign objects, chemicals, and drugs from entering the brain. The only types of substances that can make it to the brain are either neurotransmitters or drugs that mimic the molecular structure of neurotransmitters. It is for this reason that vasogenic edemas are particularly devastating. There are three different subtypes of vasogenic edema. Hydrostatic vasogenic edema occurs in patients with acute malignant hypertension. Cases of hydrostatic vasogenic brain edema are believed to be caused by conflicting pressures within the cerebral capillary network with the exuding of extracellular cerebral fluid to the other capillaries. Pressure from cancerous cells in the brain can put pressure on the brain structures and trigger vasogenic edema.

We have stated earlier that this is not the type of cerebral edema that Bruce Lee suffered from. He did not have malignant hypertension (excessively high blood pressure) and he did not have cancerous cells in this brain.

Vasogenic edema can also occur as a result of changes in pressure from being in high altitudes. It is an extreme sign of altitude sickness. High altitude cerebral edema is the potentially fatal form of altitude sickness. Symptoms of the high altitude cerebral edema include psychotic behavior, loss of coordination, headache, decreasing levels of consciousness, disorientation, memory loss and loss of coordination. We know that his death was nothing to do with this type.

Cytoxic brain edema is another type of cerebral edema that is caused by improper cellular metabolism of the glial cells in which the sodium and potassium pumps do not work properly, resulting in swollen brain cells. The blood-brain barrier is left intact in this type of cerebral edema. This condition is a sign of intoxications from dinitrophenol, triethylene, hexachlorophene, and isoniazid in cases of Reye's syndrome. In other words, from medications that are commonly used in medicine. As we have already stated, cerebral edema is treated with diuretic medications, Mannitol, or cerebral decompression surgery.[8]

And so further, after our brief review of cerebral edema, back to the issue of the 'back injury' of Bruce Lee. Other martial artists and those close to Lee state it happened.

In an interview conducted with Robert Lee, the brother of Bruce Lee, the following is stated:

COF (Interviewer): Bruce lost a great deal of body mass during 1973 and ceased his regular training six weeks prior to his death in July due to exhaustion. Did this happen?

Robert Lee: Bruce did loose a lot of weight close to his passing. I can't go into detail on why this was, at this time. I will say that the Lee family will come out with much desired information surrounding Bruce's death, soon as the right time permits. Bruce did not do himself in. There were other factors involved

COF: Can you tell us anything about your investigation regarding your brother's death? Do you know what exactly happened to Bruce with his back injury? Was it caused by lifting weights?

Robert Lee: Bruce's death has been investigated and always will be. The Lee family does believe there was foul play. In the future the Lee family will be coming out with projects that support our beliefs. There is much to say, and much did happen. When the time is right, all will see the truth. Bruce hurt his back from weights not a fight like you see in Dragon: The Bruce Lee Story.'[9]

Apart from the issue about the weights we are also given statements about 'foul play'. Again, after so many years the conspiracy theorists have carried this viewpoint around, that Lee was subject to foul play, without having done anything about it or presented the actual hard evidence. Waiting until the time is right. How many decades do we have to wait until the time is right?

So according to the interview with Robert Lee, Bruce injured his back by lifting weights. From 'Unsettled Matters' we get a different account. Lee attended the office of a Dr. Lionel Walpin in September of 1970 who examined him. Reportedly his back pain initially occurred while he was engaged in an activity quite different from lifting weights, as is detailed in the book. I will leave it to readers to read 'Unsettled Matters' regarding this. The doctor ordered X-rays but concluded that there was little wrong with him.[10] This is odd. I have no issue that this occurred. I'm just perplexed as to why the fittest man in the world, who reportedly had a high pain threshold, would go to a doctor's office and undergo an x-ray, which happened to reveal, well nothing. Or nothing pathological that is. The merry-go-round of doctors started when he then was advised, by a friend, to see a Dr. Herbert Tanney. Again, reportedly, this doctor also could find nothing seriously the matter with Lee. However, he commenced treating Lee with injections of Cortisone.[11] One authority, in reference to such procedures states, 'In recent decades, Cortisone medications have been injected into the space around the spinal cord (epidural space) to reduce the inflammation and swelling of the disc herniation, thereby relieving irritation of the adjacent nerves. It has never been certain as to whether this procedure (epidural injection) can actually reduce the need for surgery. A study by Simon Carette, MD, and others from Laval University and the University of Montreal, looking at the long-term benefits of epidural injection for sciatica from disc herniation, was published in the New England Journal of Medicine.

Dr. Carette's study found that although epidural injections for disc herniation of the low back relieved pain in the leg temporarily, the benefit was short-lived. In fact, after three months, there was no benefit from epidural injection compared to saltwater placebo injection. Further, the need for surgery was not influenced by the injection even one year later.[12] And so the Cortisone would not have been likely to have produced any long term benefit. Then, reportedly in December of 1970, Dr. Tanney referred Lee to a Dr. Ellis Silberman, who examined him and ordered another set of x-rays. Bruce Lee would have been glowing in the dark by now, with all of these x-rays. Dr. Silberman's findings were that, 'Examination of the lumbosacral spine and pelvis was within normal limits.'

And so do we conclude from this? That in fact there was no back injury?

Let's take a look at what back x-rays can and can't reveal:

'You might think that a back x-ray film would be important in understanding back pain, but back x-ray films usually do not help much. People with the usual kinds of simple backache generally have normal x-ray films. Regular x-ray films cannot recognize pressure on a nerve or the presence of spinal stenosis.'[13] Interesting. Remember, this was before the days of MRI equipment and far more sophisticated investigative procedures. Or this:

'My patient went to speak to his doctor about these spasms, and he told me an x-ray was ordered. I asked why he thought he needed an x-ray for non-specific muscle spasms that seemed to occur spontaneously. He answered that the x-ray would tell the doctor what is causing the spasms. While an x-ray (or any other diagnostic study) may help provide a diagnosis, often times these studies are used to rule something out, not in. An x-ray is great to show a fracture in a bone, but spontaneous onset of muscle spasms is not a typical symptom of a spinal fracture in a healthy man. Plus, many symptoms of back pain are non-specific-they simply cannot be attributed to a single pathology or problem.'[14] Further, X-rays cannot diagnose a herniated disc, but they can identify tumors and bone spurs. A computerized tomography (CT) scan will show impinged nerves. The most complete test is an MRI (magnetic resonance imaging), which provides clear images of your spine, soft tissues and nerves.[15] As stated, MRI's simply did not exist at the time of Lee's injury. The first image ever produced by Magnetic Resonance Imaging was produced in July of 1977.

If still in doubt, this may clarify matters:

'X-ray imaging includes conventional and enhanced methods that can help diagnose the cause and site of back pain. A conventional x-ray, often the first imaging technique used, looks for broken bones or an injured vertebra. A technician passes a concentrated beam of low-dose ionized radiation through the back and takes pictures that, within minutes, clearly show the bony structure and any vertebral misalignment or fractures. *Tissue masses such as injured muscles and ligaments or painful conditions such as a bulging disc are not visible on conventional x-rays.* This fast, noninvasive, painless procedure is usually performed in a doctor's office or at a clinic.[16] (Author's italics)

And so it is clear that even though Lee's x-rays demonstrated no discernable pathology that is not to say that there was not any actual pathology there. Someone can still experience significant back pain even with normal spinal x-rays if there is inflammation, or pressure on a nerve, for instance. Apparently Bruce Lee continued to return to Dr. Tanney's office on a semi-regular basis for Cortisone (Depo-medrol) injections.[17] And that begs the question, why did the doctor continue to administer medications if in fact there was nothing wrong with him? Why not give him a placebo? Or why not just say, "Mr. Lee there is nothing wrong with you, you need to see a psychiatrist because you are suffering from a psychosomatic condition?"

Bruce Lee was a highly intelligent man, and it is unlikely that he would have gone to a doctor's office complaining of back pain if this was not the case. What would have been the gain? And, as we have demonstrated by doing a bit of research, a person can experience significant back pain and back or spinal problems without x-rays detecting this. So it is likely, in my opinion, that he did indeed suffer from back pain and from a back injury. Formal medical records that Mr. Bleecker has or has seen cannot be contradicted. But sometimes they may simply not tell the whole story, and just because nothing showed up on x-rays does not mean that he did not have an injury or some pathology. The common Bruce Lee fan, like me, is then left in the dark as to how it occurred. Was it indeed suffered while he engaged in other activities? Was it aggravated or did he have a relapse caused by lifting weights incorrectly? Possibly. So there is no reason at all why Bruce Lee could not have suffered from chronic back pain, even though the x-rays were negative. Now, the Bruce Lee Foundation states, in its biography of Bruce Lee, that 'After much pain and many tests, it was determined that he had sustained an injury to the fourth sacral nerve.' So if, as they claim, he had injured his fourth sacral nerve, then we would not expect that to show up on an x-ray anyway, because spinal x-rays, as we have determined above, will only show up bone structures, and not soft tissue or nerves or the inflamed or damaged nerves. In all fairness to the Bruce Lee Foundation, that is all that they are saying. Not that he had any *structural* injury to the spinal vertebrae, which would have shown up on x-rays. Some may work under the assumption that x-rays of the spine will detect *all* pathology and *all* of the causes of back pain, which they patently, unequivocally and simply do not. The further issue is; how extensive, how crippling and how significant was this back injury? Here we can really get into dispute and conjecture, because out there just about every person has an opinion on this. From Lee being mildly impaired, to being bed ridden for six months, and needing on-going medication and treatment. I would encourage people to do their own research and make up their own minds. My own opinion is that Bruce Lee suffered from chronic back pain. Bruce Lee would not in my view have attended a doctor's office complaining of back pain if this was not the case. However I do not believe that Bruce Lee was incapacitated to any significant extent by it. Mr. Bleecker states that subsequent to the back injury that Bruce Lee reported to his doctor there are pictures of him performing high kicks.[18] If this is the case then I bow to Mr. Bleecker's evidence. Anyone with a *serious* spinal injury would not be able to perform high kicks. But that does not mean that Bruce Lee did not have a back injury, or at least a problem with his fourth sacral nerve that caused him pain. This in turn led him to taking medication to control this pain and the inflammation of the nerve; symptoms that we assume were present. However it was not disabling, and it some state clearly that Lee was not laid up for months where he wrote significant and extensive martial arts tomes and all of the other depictions of this phase of his life. As a Bruce Lee fan I simply want to know the truth. Just to cement my arguments prior to closing let me give you one more example from the medical literature:

'Chronic back pain is, in some ways, a loose construct. In many instances, physicians do not know why a person has back pain. X-rays and other imaging techniques often do not show any major abnormalities that might cause the pain. In many other instances physicians cannot explain why a person does not have pain when their x-rays show significant problems. In addition, when there is a detectable abnormality of the spine, the amount of pain an individual feels is not correlated to the medical severity of the spinal problem. Very significant problems may cause not much pain at all, and minor problems may be extremely painful.'[19]

So it is important not to insinuate that just because Lee's x-rays of his spine were negative that there was no injury present. In any case, reportedly since September of 1970, Lee had become a frequent visitor to the office of Dr.Tanney for the cortisone injections. In July, 1971, Bruce Lee traveled to Thailand in order to appear in the movie 'The Big Boss'. He reportedly, '....began losing weight, which was most likely the result of his abrupt discontinuance of Cortisone.'[20] Lee had also reported that the Thai village in which he was in had food available that was, in Lee's words, 'terrible', and that there was no beef and very little chicken. He stated that he was glad he had brought his vitamins. Perhaps his loss of weight was due to the fact that he wasn't eating properly, and obviously getting next to no protein. In an athlete like Lee, where there is not adequate protein intake, this can lead to a catabolic state. A catabolic state is a condition that is mainly caused by excessive training coupled with a lack of adequate nutrition, especially protein. It results in numerous undesirable side effects in the body, such as extreme fatigue, joint and muscle pain, and sleeplessness.[21] Sometimes, the most simple of explanations will suffice. Perhaps his loss of weight had nothing to do with Cortisone at all.

Chapter 5
Illicit drug use – a factor in his death?

Illicit drug use is a sensitive and controversial topic. I do not under any circumstances advocate the use of illicit drugs, however there is one thing that must be said, loud and clear. That is, that prescribed medications kill far more people than illicit drugs, and that includes the illicit drug such as marijuana, which Bruce Lee happened to be using at the time of his death. Let's view the evidence in support of my statement:

'Nearly 20 percent of Americans have used prescription drugs for nonmedicinal reasons, three-quarters of whom may be abusing them. Legal prescription drug abuse is a silent epidemic, and is part of the reason why the modern American medical system has become the leading cause of death and injury in the United States. Authored in two parts by Gary Null, PhD, Carolyn Dean, MD ND, Martin Feldman, MD, Debora Rasio, MD, and Dorothy Smith, PhD, the comprehensive Death by Medicine article described in excruciating detail how everything from medical errors to adverse drug reactions to unnecessary procedures caused more harm than good. Seven years after the original article was written, an analysis in the New England Journal of Medicine, November 25, 2010.... found that, despite efforts to improve patient safety in the past few years, the health care system hasn't changed much at all. In a June 2010 report in the Journal of General Internal Medicine, study authors said that in looking over records that spanned from 1976 to 2006 (the most recent year available) they found that, of 62 million death certificates, almost a quarter-million deaths were coded as having occurred in a hospital setting due to medication errors.

An estimated 450,000 preventable medication-related adverse events occur in the U.S. every year.

The costs of adverse drug reactions to society are more than $136 billion annually - greater than the total cost of cardiovascular or diabetic care.

Adverse drug reactions cause injuries or death in 1 of 5 hospital patients.

The reason there are so many adverse drug events in the U.S. is because so many drugs are used and prescribed – *and many patients receive multiple prescriptions at varying strengths, some of which may counteract each other or cause more severe reactions when combined.*[1] (Author's italics)

Please keep note of the above, because when we discuss the death of Bruce Lee it is significant. Adverse drug reactions can happen to any individual, but there are variations in the effects of drugs too due to racial factors for instance. And individuals can and do experience adverse drug reactions, even severe or fatal reactions, when they have used the same medications previously without any problems.

For centuries, people of different cultures have used opiates to relieve pain. Now, it appears that sensitivity to the opiate codeine varies with ethnic background, according to a recent study. These findings could help doctors treat pain more effectively in different individuals. Codeine's analgesic properties stem mainly from the body's ability to metabolize it into morphine, a much more potent opiate, says Alastair J.J. Wood of Vanderbilt University School of Medicine in Nashville. His research indicates that further reactions contribute to the painkilling response. Wood and his colleagues examined the effects of codeine in men of European extraction and in Asian men. Both groups transformed codeine into morphine similarly, but the Asian men experienced significantly weaker effects from the drug. Wood presented the findings this week at a meeting of the American Chemical Society in Las Vegas. "It's a nice piece of work," says Wendel L. Nelson of the University of Washington in Seattle. Researchers had suspected that morphine is responsible for the pain relief provided by codeine, but the current study "is the clinical piece that really nails it down." Previous studies have shown that some people lack an enzyme called CYP2D6 that chemically alters codeine into morphine. The same enzyme metabolizes many drugs used to treat high blood pressure, heart arrhythmias, and depression. About 8 percent of whites, 6 percent of blacks, and 1 percent of Asians do not produce CYP2D6. Currently, doctors tend to think that patients who don't respond to a painkilling drug need higher doses. "Here's an example where a portion of the population will get no effect, and increasing their dose a lot more will still produce no effect," Wood says. "It's not that they're wimps or their pain is worse than [that of] other people." All of the people in the current study, 10 white men from the United States and 8 men from China, possessed CYP2D6. Pain control is difficult to measure, so the researchers monitored how codeine affected breathing, blood pressure, and pupil dilation. Consistently, codeine affected the Chinese men less than the U.S. men. Their bodies cleared morphine faster and increased metabolism of codeine through enzymes other than CYP2D6."[2] These are the racial factors that we must take into account when discussing medications and their effects, and what constitutes an appropriate or effective dose of medication.

Whenever we discuss issues in regards to substance use, substance addiction and substance tolerance we have to remember these racial factors that come into play and can make a significant difference in terms of tolerance to the drug or specific reaction to the drug. We know that Asians have a very difficult time tolerating alcohol. The tolerance to alcohol is not equally distributed throughout the world's population, and genetics of alcohol dehydrogenase indicate resistance has arisen independently in different ethnic groups. People of European descent on average have a high alcohol tolerance and are less likely to develop alcoholism compared to Aboriginal Australians, Native Americans and some East Asian groups. This is related to an average higher body mass, but also to the prevalence of high levels of alcohol dehydrogenase in the population. The high alcohol tolerance in

Europeans and some other ethnic groups has probably evolved as a consequence of centuries of exposure to alcohol in established agricultural societies. Not all differences in tolerance can be traced to biochemistry. Differences in tolerance levels are also influenced by socio-economic and cultural difference including diet, average body weight and patterns of consumption. An estimated one out of three people in East Asian countries have an alcohol flush reaction, colloquially known as "Asian Glow", a condition where the body cannot break down ingested alcohol completely because it lacks the genetically coded enzyme that performs this function in the bodies of drinkers with "European" tolerance levels. Flushing, or blushing, is associated with the erythema (reddening caused by dilation of capillaries) of the face, neck, and shoulder, after consumption of alcohol.[3] And so it is likely that Lee would have had a difficult time handling alcohol.

In regards to Lee's marijuana use we have the following comment:

'Also, Bruce had side effects from Nepali hashish - severe headaches. Even the movie Dragon: The Bruce Lee Story and all his close friends would say Bruce had severe headaches. Go and speak with the Chinese elders who would be the same age as Bruce Lee would have been today - in their 60's and 70's and tell them about *"Dai-ma"* pronounced *"Die-Mah"*. It's the Cantonese word for Hashish and they will also agree that is what killed him. It's worth noting that to understand what killed Bruce is to know what type of symptoms he had which led to his death: 1) vomiting 2) seizures 3) severe headaches, fainting and coma. All these symptoms run hand in hand with herbal poisoning. In fact, if you do a [internet] search 'vomiting + seizures + fainting/and or coma' you will get a long list of results showing 'poisonous herbs.' The writer goes on to note that Bruce Lee had very low body fat, 'zero or very low body fat. This body fat would help buffer the potent effects of Nepali hashish. All the respected martial arts authorities...... know that it was hashish that killed him.'[4] Let's engage in a rational rebuttal of these statements and this is easily done because medically, vomiting, seizures, fainting and/or coma can be caused by a huge number of conditions, poisoning by herbs being only one of them. As an aside to this post, the other hypothesis that has been circulating relates to the fact that Lee had the sweat glands under his armpits removed after consulting a doctor in Hong Kong in November of 1972. Lee had complained to this doctor of weight loss, sunken cheeks, acne and profuse sweating.5 The doctor, in regards to the sweating complaint, diagnosed axillary hyperhidrosis. That is simply excessive sweating in the armpits. Lee reportedly was later admitted to hospital and underwent surgery to remove his axillary sweat glands. The procedure itself takes 1 to 1 1/2 hours and is done with the patient anaesthetized. Two small incisions are made in the underarm area and a special curette is used to scrape away the sweat glands that reside just on the undersurface of the underarm skin.6 The 'death of Bruce Lee was caused by hashish' theorists jumped on this straight away and I read one post to a website which stated that Lee died because his body couldn't 'sweat out' the toxins from the hashish. Again there is a perfectly logical and rational rebuttal to this. Axillary sweat gland removal does not affect sweating in any other part of the body. If it did the individual would die from hyperthermia as sweating is a necessary thermoregulatory mechanism. And just to nail this one on the head completely, so that we are not subject to these same assertions about 'sweating out toxins' time and time again, an article that I read as part of my research states:

'The bottom line: Sweat does contain trace amounts of toxins, says Dr. Dee Anna Glaser, a professor of dermatology at St. Louis University and founding member of the International Hyperhidrosis Society, a medical group dedicated to the study and treatment of heavy sweating. But, Glaser, adds, in the big picture, sweat has only one function: Cooling you down when you overheat. "Sweating for the sake of sweating has no benefits," she says. "Sweating heavily is not going to release a lot of toxins." In fact, Glaser says, heavy sweating can impair your body's natural detoxification system. As she explains, the liver and kidneys - not the sweat glands - are the organs we count on to filter toxins from our blood. If you don't drink enough water to compensate for a good sweat, dehydration could stress the kidneys and keep them from doing their job. "If you're not careful, heavy sweating can be a bad thing," she says.[7]

'His doctors knew what almost killed him on May 10th, 1973. He even presented it to Dr. Langford, the next day, when Langford came in and interrogated him at the hospital, after saving his life. He wanted to know if Bruce was taking any drugs. There was no other explanation. He pulled out Nepal hashish and Langford told if he started taking this again, it would kill him. He rejected Langford's medical advice and flew to UCLA and was clean. They found nothing wrong with him, because there were no drugs in his system.'[9] And interestingly, no withdrawal symptoms either! The 'death due to hashish' theorists continue, 'This gave him a false sense of security and a few weeks before he died, he was ingesting it again. This type of hashish he got from Katmandu, Nepal. It's one of the most near-lethal strains of 'unrefined' hashish and is much more rare (sic) than the refined types manufactured in the Middle East. It's been documented to kill its users the exact same way it killed Bruce on July 20th, 1973. It's neurological side effects are nightmarish, and contains over 4,000 chemical compounds, any one of which, his central nervous system could build up a sensitivity to, and so highly toxic, cerebral edema, kidney/ adrenal failure, vomiting, convulsions, cardiac arrest, coma and death is the result.'[9] There are some problems that I have with the above statement. One is, individuals can make any statements they wish however when the statement runs, 'It's been documented to kill its users the exact same way it killed Bruce on July 20th, 1973.' I simply wish that they would source and reference this information. Where is this documentation that they refer to? And was the substance he was using at the time subject to laboratory analysis and its chemical constituents precisely defined? The substance he was using at the time was never as far as I am aware subject to analysis so we don't actually know its precise constituents. It is conjecture and speculation. Further, let me give you some information:

'Cigarette smoke contains over 4,000 chemicals, including 43 known cancer-causing (carcinogenic) compounds and 400 other toxins. These include nicotine, tar, and carbon monoxide, as well as formaldehyde, ammonia, hydrogen cyanide, arsenic, and DDT. Nicotine is highly addictive. Smoke containing nicotine is inhaled into the lungs, and the nicotine reaches your brain in just six seconds. While not as serious as heroin addiction, addiction to nicotine also poses very serious health risks in the long run.'[10] The plain old cigarette has been used for centuries and people smoke for decades only to succumb to the toxic effects of it after many years. It's a slow death, not a sudden death. Even with all of those toxins the body copes, for a while at least. One doesn't smoke cigarettes and suffer, quote, '....cerebral edema, kidney/ adrenal failure, vomiting, convulsions, cardiac arrest, coma and death is the result.' Despite the more than 4,000 chemicals in a cigarette. Why should that happen with marijuana also, even if it is 'tainted'? The fact is that it's a slow and progressive damage to one's health that is sustained by cigarette smoking. Even if we take on face value the premise and assumption that the marijuana that Lee was using had toxins in it, having toxins as a constituent does not of itself equate with sudden death. Cigarettes are full of them and people are happily puffing away every day without incident. But the habit will slowly lead to an early grave.

In contrast to the writer's assertion that I document above, that there are over 4,000 chemical compounds that are in marijuana other literature, from the Americans for Safe Access organization, for example states:

'...[T]here are 483 different identifiable chemical constituents known to exist in cannabis. The most distinctive and specific class of compounds are the cannabinoids (66 known), that are only known to exist in the cannabis plant.

Other constituents of the cannabis plant are: nitrogenous compounds (27 known), amino acids (18), proteins (3), glycoproteins (6), enzymes (2), sugars and related compounds (34), hydrocarbons (50), simple alcohols (7), aldehydes (13), ketones (13), simple acids (21), fatty acids (22), simple esters (12), lactones (1), steroids (11), terpenes (120), non-cannabinoid phenols (25), flavonoids (21), vitamins (1) [Vitamin A], pigments (2), and elements (9). The very most of these compounds are found in other plants and animals and are not of pharmacological relevance with regard to the effects exerted by cannabis preparations.' Even steroids are present too. Interesting.

The second issue that I have is that Bruce Lee was being characterized as something of a drug addict. However, there are really two objections to this. One is that if he were so addicted to drugs why did he leave his source of supply and return to America to undergo tests at an American hospital after the May 10 incident, '….he flew to UCLA and was clean.' Drug addicts would not be too keen on leaving their source of supply, if they are addicted. You can imagine the conversation with someone who is stoned, 'Fly to the States? No way, man. Hell my dealers here and I've got my women lined up and my can of hashish and my dealer's flush with all kinds of gear. If I get checked out there by a doctor he'll either see that I'm stoned or going cold turkey. No way, man, I'm keeping my shit and my ass here!' That's not referenced by the way because I made it up. Let's move on. Secondly, but seriously, if he were addicted then if he was clean when he 'flew to UCLA' *why was he not experiencing the symptoms of withdrawal from the drugs that he was allegedly addicted to?* There is no doubt that Bruce Lee used marijuana. It was present in this stomach and small intestine at autopsy. However it was present in relatively small quantities. We know that Lee had only been drinking soda drinks on the day of his death. The evidence is clear, he had ingested marijuana. The amount that he had ingested, as Professor Teare pointed out at the inquest into Lee's death, was simply not significant. And there is no evidence that it caused his death, or anyone else's who was around him at the time, and who would have been using the same substance.

Professor Teare knew a lot about drugs, and he was right when he said that the marijuana in Lee's stomach was insignificant. It has been stated that, 'Tetrahydrocannabinol (THC or the active ingredient in marijuana) is a very safe drug. Laboratory animals (rats, mice, dogs, monkeys) can tolerate doses of up to 1,000 mg/kg (milligrams per kilogram). This would be equivalent to a 70 kg person swallowing 70 grams of the drug - about 5,000 times more than is required to produce a high. Despite the widespread illicit use of cannabis there are very few if any instances of people dying from an overdose. In Britain, official government statistics listed five deaths from cannabis in the period 1993-1995 but on closer examination these proved to have been deaths due to inhalation of vomit that could not be directly attributed to cannabis (House of Lords Report, 1998). By comparison with other commonly used recreational drugs these statistics are impressive.'[11] The drug also has antiemetic properties, that is it suppresses nausea and vomiting. One report, in regards to the toxicity of hashish states, *'Death from acute poisoning is extremely rare, and recovery has occurred after enormous doses.'* 12 (Author's italics) The other issue that must be stated is that usually ingesting such drugs occurs in the context of a social network and would lead us to ask the very important question, *who else used the same drug with Bruce Lee?* Did he share his hashish and did others share their stash with him? And if he did share it and use it in the presence of others, then why, if it was so toxic, is everyone else that surrounded him who also used it alive and he is dead? Was there a cluster of deaths due to the use of the same drug that Bruce Lee was using in Hong Kong at that time? Were all of the others that were using marijuana with Bruce Lee or around him, or in Hong Kong in general, dying, falling ill, or experiencing episodes of cerebral edema or just getting plain sick due to the 'toxins' in it? If they were, why was it not reported? And also, Professor Donald Teare was a very smart man with enormous experience in these matters, and he did not feel that the illicit drug in Bruce Lee's stomach was in any way related to his death at all.

The fact of the matter is that on the night of his death, Bruce Lee may simply have had a severe headache, of the kind any person can get any day. For some reason he was not immediately taken to a doctor, and nor was a doctor called at the point in time in which this occurred. And we can only speculate as to its cause. Was the headache symptomatic of the early stages of cerebral edema? Or was it an 'ordinary' headache and did the cerebral edema occur after he took the Equagesic in reaction to it? I don't know, Tom

Bleecker does not know, and no one else knows the answer to this, because Bruce Lee was not medically examined at the time. Mr. Bleecker paints the scenario in his book that the cerebral edema *preceded* the taking of the Equagesic tablet. This is conjecture because there was no doctor present at the time who was able to examine Lee and to determine if the headache was due to a developing cerebral edema, or was just a simple headache. It's possible, but just as likely to be the other point of conjecture that the headache was not the result of cerebral edema, it was just an ordinary headache, and the cerebral edema occurred in response to the Equagesic tablet. No one knows exactly, and we never will know. The Equagesic hypothesis is simply the most probable because there was no other evidence of any other toxic substance being present in his body. He had a headache. Not as a result of ingesting hashish and not as a result of anything else apart from the incredible stress that he was under at the time. *If* Betty Ting was using the same substance at the same time, and I don't know if she was or she wasn't. But let's say they shared their marijuana. It seems to be a social thing from what I understand of the scene. *If* Betty Ting used the same marijuana that Lee used, and it was so toxic, why didn't she become ill? The following brief biography of Ms.Ting Pei is as follows:

'Betty Ting Pei was born in Beijing, China to a prominent and politically influential family. Her father and grandfather were physicians, and her mother was descended from warlords. The family migrated to Taiwan when she was two. In 1968, she joined the Shaw Brothers studio in Hong Kong, and Betty was one of its rising starlets when in 1971 she met martial arts superstar Bruce Lee. The controversy and many rumors surrounding Lee's mysterious death in her apartment in 1973 seriously damaged her own career. Although insisting for years that she and Lee were just friends, she admitted in a 2006 radio interview that they had been lovers for more than a year at the time of his death. She also said that the public censure and criticism she received had driven her to drug use as a means of escape, and while she was now clean, she still had psychological problems.'[13]

THE DEATH OF BRUCE LEE: A CLINICAL INVESTIGATION

The biography above reports on '....Lee's mysterious death in her apartment...' In fact there was nothing mysterious about it at all. He died from an unforeseen adverse drug reaction to the medication which had been given to him by Betty Ting. Such lethal reactions to medications happen all over the globe every day. And there was nothing mysterious as to why he was at her apartment in the first place. Lee was given and ingested an Equagesic tablet, which was notorious for its toxic side effects, one of which is clearly documented in the medical literature: cerebral edema. There was no doctor present at the time that he complained of his headache so he was not examined at that immediate point in time. If he had been examined there could have been a conclusive determination made if his headache was due to increased intracranial pressure as a result of cerebral edema, or if it was a common or garden headache, or a migraine, perhaps due to stress.

For those of you who still have some doubts about the safety of marijuana, read this:

'Marijuana has been used as a medicinal herb for thousands of years, going back to ancient civilizations in Egypt, India and Africa. In all that time, up to and including the present day, there has never been a report of a fatality directly due to the consumption of marijuana. In contrast, over 1,000 people die annually in the US from an overdose of our most common non-prescription drug, aspirin. In addition, many thousands of deaths result from the legal prescription drugs.'[14] Or this:

'Drugs used in medicine are routinely given what is called an LD-50. The LD-50 rating indicates at what dosage 50% of test animals receiving a drug will die as a result of drug induced toxicity. At present it is estimated that marijuana's LD-50 is around 1:20,000 or 1:40,000. In layman terms this means in order to induce death, a smoker would have to consume 20,000 to 40,000 times as much marijuana as is contained in one marijuana cigarette. NIDA-supplied [National Institute of Drug Abuse] marijuana cigarettes weigh approximately 0.9 grams. A smoker would have to consume nearly 1,500 pounds of marijuana within about 15 minutes to induce a lethal response. In practical terms, marijuana cannot induce a lethal response as a result of drug-related toxicity.'[15]

Or further:

'No acute lethal overdoses of cannabis are known, in contrast to several of its illegal (for example, cocaine) and legal (for example, alcohol, aspirin, and acetaminophen) counterparts.'[16]

However, and I must document this in order to engage in objective and balanced research, we do have in the literature at least one contrary view:

'Each of the 3 cannabis-associated cases of cerebellar infarction was confirmed by biopsy (1 case) or necropsy (2 cases)... Brainstem compromise caused by cerebellar and cerebral edema led to death in the 2 fatal cases.'[17]

Let's review this information. I went to the source of the above statement, the article itself. In it it states, 'From the literature, it is clear that marijuana use can cause systemic hypotension, impair peripheral vasomotor reflexes, and may alter central nervous system blood flow and cerebral vascular autoregulation. Marijuana use has been associated with *stroke* in adults, but acute central nervous system infarction related to marijuana use is not well described in children. Although the mechanism of neurologic injury and its localization to the posterior circulation in these cases remains uncertain, our observation of *acute cerebellar infarction* in 3 adolescents shortly after marijuana use suggests that this drug may contribute to cerebellar vascular injury, possibly by causing vasospasm, especially in the inexperienced or episodic user, resulting in cerebellar ischemia. The cerebellum may be more susceptible to ischemia, because it is a region without rich collateral circulation.' (Author's italics) And so we are discussing here cerebral edema *as a consequence* of acute cerebellar infarction. In other words, a stroke or cerebrovascular accident as it is otherwise referred to. The medical journal article that I source this from reports on the autopsy of one of the patients, 'At autopsy, the 1660-gram brain (age-matched reference range: 1416–1516 gram) showed moderate diffuse cerebral edema with gyral flattening, sulcal narrowing, and compression of the lateral ventricles. The left cerebellum showed asymmetric tonsillar herniation and an acute infarct. The infarct contained foci of acute hemorrhage, although no hematoma was present. Histologically, the cerebellar infarct showed focal hemorrhage and infiltration by neutrophils and macrophages (histologic features of an infarct ~3–7 days old.)' However, we must state that on autopsy Bruce Lee's brain showed cerebral edema, *however there was no cerebrovascular event such as an infarction.* (An infarction is defined as, 'A localized area of dead tissue (necrosis) resulting from obstruction of the blood supply to that part, especially by an embolus.')[18]

To conclude, the overwhelming amount of research and evidence points to the fact that marijuana is a relatively safe drug, far safer in fact than many prescription medications.

Let's take a look now at the issue of Lee's weight, which has been subject to discussion and conjecture as to why it fluctuated. 'Bruce's weight in 1972 was 146 pounds, as he stated in a phone-taped interview with Alex Ben Block, who called him while he was making Way [of the Dragon] in summer 1972. He looks about the same weight in the filmed footage GOD [Game of Death] some 3 months later in Sep-Oct the same year. Healthy weight. When he arrived in HK in 1970, he was 155 pounds, and weight training trimmed him down. In 1973, while making ETD [Enter the Dragon], he was about 135 pounds and underweight. By May, he was down to 126, some 20 pounds. I've seen a few photos of Bruce shortly before his death and he looks frail and sickly. The robustness from his face is gone. Dr. Langford said he looked obscene, with only 1% of body fat on him.'[19] Well, this may be so, but again we have the apparent historical record of him after his first episode of cerebral edema going to UCLA to undergo a medical examination and no doubt an extensive battery of blood tests. To put his weight in perspective, his height was recorded as 5ft 8. Some say 5ft 7. He has written his height as 5ft 8 before he was famous and 5ft 7.5 has been mentioned in some of his books. His Brother Robert Lee said on the Early Years website 'Bruce was an inch taller than I. I am 5/7 and Bruce was 5/8 with a shoe size of 7-71/2. I know it's a little confusing, because we have all heard for years in how small he was, but in fact 5/8 is really not that small.'[20]

In terms of the 'ideal' weight, there is no specific figure, but a range. The following illustrates this:

'The Centers for Disease Control and Prevention has a chart that informs men what their ideal weight is if they have a BMI between 18.5 and 25.

If you're a 6-foot man, your ideal weight is 137 to 185 lbs. Other ideal weights in pounds, with heights in parenthesis, include 101 to 135 (5-2), 104 to 140 (5-3), 108 to 145 (5-4), 111 to 150 (5-5), 114 to 155 (5-6), 117 to 160 (5-7), 121 to 165 (5-8), 125 to 170 (5-9), 129 to 175 (5-10), 133 to 180 (5-11), 141 to 190 (6-1), 145 to 195 (6-2), 149 to 200 (6-3) and 153 to 205 (6-4).'[21]

So, if Lee was between 5'7" and 5'8" (to complicate matters individuals are usually taller in the morning after sleep due to compression of the spinal structures that occurs during the day due to being in an upright position.) then his weight should have been in the 117lbs to 165lbs range. This issue of the body fat is important and 1% is by any standards far too low, especially for an athlete.

After the May 10 incident, 'A week later Lee, feeling fit as ever, was examined in Los Angeles by Dr. David Reisbord, who did a brain scan, a brain flow study, an EEG, and a complete physical. The doctor concluded that Lee had suffered a grand mal seizure of no known origin, but that overall *he was in extraordinarily good health.*'[22] (Author's italics)

What did the doctor's there say after examining him at that time? Was it, "Oh my gosh, this man is in a bad way, we had better admit him immediately. He's got symptoms of drug withdrawal, he's emaciated, he's very sick, he's exhausted, he's suffering alcohol withdrawal? He's got this in his blood, he's got that in his blood, these levels of (whatever) are way over the top, or these blood levels of (whatever) are way down. Any of that? No! He was given a 'clean bill of health' and then returned to Hong Kong. Only being prescribed an anti-convulsant, to prevent seizures of the like he experienced on May 10. An anti-convulsant, which apparently was not present in his body at the time of his death. So contrary to any myth that is out there, that he was gobbling every prescription and illicit drug in sight, as was reported while he was on location in

Thailand and later, there was little to no direct evidence of that. On the night that of his death he hadn't even taken the Dilantin anticonvulsant medication that had been prescribed to him. If he had presented to UCLA in any state of health that was in any way markedly deviant from what we consider normal, the doctors there would have picked it up straight away.

I can't speak for Tom Bleecker but from what I understand one of the motivations that he had in writing 'Unsettled Matters' was to expose how an obsessive drive for perfection can lead to disastrous consequences, and if this is so, Tom Bleecker and I can at least agree on this. Bruce Lee was an obsessively driven man. He sought perfection of the physical and psychological self. Reportedly, at various times in his life he resorted to liquid diets. This has variously been stated as allowing him, during his Kung Fu movies, to stay longer in the air with his flying kicks, such was the litheness that his body displayed. So the reasons for his weight loss are no doubt multifactorial, and may have related in part in Hong Kong to the stress that he was under. This of itself can lead to disastrous consequences, along with steroid use and diuretic use, if this was occurring, in anyone.

The obsessive drive for physical perfection, a drive that we may postulate and hypothesize was present in Lee due to the fact that he viewed himself, due to his cryptorchidism as somehow imperfect, can be evident in society in general, albeit for other reasons. The obsession in modern-day society with the body, and a slim body, leads many females especially to engage in various fad diets and vitamin regimens. I do not know the precise vitamin regimen Lee was using at the time of his death or just before, but reportedly often he would only drink carrot juice and orange juice, as a substitute for other food. High doses of certain vitamins can have significant adverse effects, and individuals can suffer toxic effects from various natural substances that they take without realizing the harm they are doing to themselves.

This is one other aspect of his life and behavior that I wish to consider. We know that Bruce Lee was obsessive in regards to his training regimen. He was a perfectionist and that was precisely why he was able to do what he did. However in individuals with these personality traits they can be driven to engage in behaviors which, although they seek perfection, are actually maladaptive and lead to destruction as we have stated. There are also certain dangers involved with the overuse of vitamins, which he may have relied on to the exclusion of a balanced diet. Hypervitaminosis A occurs when the maximum limit for liver stores of retinoids is exceeded. The excess vitamin A enters the circulation causing systemic toxicity. Betacarotene, a precursor form of vitamin A typical of vegetable sources such as carrots, is selectively converted into retinoids, so it does not cause toxicity; however, overconsumption can cause carotenosis, a benign condition in which the skin turns orange. Although hypervitaminosis A can occur when large amounts of liver, including cod liver oil and other fish oils, are regularly consumed, most cases of vitamin A toxicity result from an excess intake of vitamin A in the form of vitamin supplements. Toxic symptoms can also arise after consuming very large amounts of preformed vitamin A over a short period of time. The U.S. Institute of Medicine says that the Lowest Observed Adverse Effect Level (LOAEL) for vitamin A, when taken over an extended period of time is 21,600 IU. Most multivitamins contain vitamin A doses below 10,000 IU, therefore multi-vitamins are unlikely to cause vitamin A toxicity when taken at their recommended dosages. But in high doses, its central nervous system toxicity can be enhanced by its lipid solubility because it is readily transported across the blood brain barrier and concentrated in the brain. Vitamin A causes cells to swell with fluid; too much vitamin A causes them to rupture in hyposmotic environments, hence the toxicity. Toxicity has been shown to be mitigated through vitamin E (tocopherol), cholesterol, zinc, taurine, and calcium. Cholesterol has been shown to prevent retinol induced golgi fragmentation.[23]

In a very interesting article that I read as part of my research Davis Miller states,

'He wasn't on a liquid diet but he wasn't eating well at all. I've talked with a couple of very capable, well-regarded doctors about this. He may very well have had an eating disorder. Again, the guy was not at all well. And he was an ultra-sensitive man. He was bigtime strung-out. As far as Tom Bleecker's assertion that Bruce Lee was a heavy drinker, here's the truth: He drank saki while living in Hong Kong. And he had very low tolerance not only for alcohol but for almost anything he introduced into his system. He got drunk very easily. And he was trying to live like Mr. Hollywood.'[24]

I have already mentioned the reasons as to why Lee would have a low tolerance for alcohol. Did Lee suffer from an eating disorder? That is a very interesting question. It is one thing to be in a remote Thai village and have little available choice of protein, but to be living in Hong Kong, which is regarded as a food haven, and not to eat properly is another. Did he over-rely on vitamins and supplements? To show what can occur when vitamin A is taken in large amounts let us turn to the medical literature.

In one medical journal a report states that three girls aged 14, 15, and 16 years had diplopia, papilledema, and other symptoms suggesting a brain tumor. In addition to various individual complaints, each had hypomenorrhea, alopecia, [loss of hair] and rhagades [linear cracks or fissures in the skin occurring especially at the angles of the mouth or about the anus] or other marked forms of dermatosis. They had been taking, respectively, 200,000; 200,000, and 90,000 units of vitamin A per day. Treatment consisted in stopping the intake of vitamin A; in addition the oldest patient received acetazolamide as a diuretic. The symptoms subsided rapidly, but in the oldest patient slight blurring of the margins of the optic disks were still present six months later. Alopecia, rhagades, nonspecific dermatitis, [inflammation of the skin] migratory arthralgia, [joint pain] hepatosplenomegaly, [enlargement of the liver and spleen] and hypomenorrhea [diminution of menstrual flow or duration] are important in the differential diagnosis between the *pseudotumor cerebri* of chronic vitamin A intoxication and true intracranial tumor.[25]

Revision of earlier estimates of daily human requirements of vitamin A has been suggested; the suggestion is that estimates ought to be *revised downwards*. Concerns exist about the teratogenicity of vitamin A, that is the capability of producing fetal malformation. The recommended daily allowance for vitamin A is 5000 international units (IU) for adults and 8000 IU for pregnant or lactating women. Being fat-soluble, vitamin A is stored to a variable degree in the body, making it more likely to cause toxicity when taken in excess amounts. In contrast, water-soluble vitamins are generally excreted in the urine and stored only to a limited extent; hence, adverse effects occur only when extremely large amounts are taken. Mortality is rare from vitamin A toxicity, although the risk may increase with higher doses.

In acute vitamin A toxicity, a history of some or all of the following may be present:

Nausea
Vomiting
Anorexia
Irritability
Drowsiness
Altered mental status
Abdominal pain
Blurred vision
Headache
Muscle pain with weakness

In chronic vitamin A toxicity, a history of some or all of the following may be present:

Anorexia
Hair loss
Dryness of mucus membranes
Fissures of the lips
Pruritus
Fever
Headache
Insomnia
Fatigue

Irritability
Weight loss
Bone fracture
Anemia
Bone and joint pains
Diarrhea
Menstrual abnormalities
Epistaxis

Carotenemia, the ingestion of excessive amounts of vitamin A precursors in food, mainly carrots, is manifested by a yellow-orange coloring of the skin, primarily the palms of the hands and the soles of the feet. It differs from jaundice in that the sclerae remain white. Isotretinoin (Accutane), a drug used for the treatment of severe forms of acne, is closely related to the chemical structure of vitamin A. The pharmacology and toxicology of these 2 compounds are similar. Birth defects if taken during pregnancy, intracranial hypertension, depression, and suicidal ideation have been reported with isotretinoin. A careful drug history to uncover this possibility is important in patients presenting with manifestations suggestive of vitamin A intoxication. Causes of vitamin A toxicity are generally categorized into acute and chronic. Acute toxicity occurs within a few hours or days after a very large intake as a result of accidental over ingestion or inappropriate therapy. The estimated toxic dose is about 25,000 IU/kg. Chronic toxicity appears after ingestion of 25,000 IU or more daily for prolonged periods.[26]

The kind of irrational dieting that Lee reportedly embarked on at times could lead to significant health problems and even to death, as has occurred in certain instances.[27] Rapid weight loss is a form of dieting that causes the body to shed more than 2 lbs. a week. They're usually highly restrictive diets, limiting your caloric intake well below the recommended levels of 1,200 for women and 1,500 for men, as indicated by the National Institutes of Health. When observed without the supervision of a health-care provider, this form of weight loss increases the risk of certain health complications. Another potential health risk of rapid weight loss is of course malnutrition, which is as we know a medical condition linked to a lack of nutrients. Crash diets and low-calorie diets can restrict eating habits to the point where you aren't getting enough essential vitamins and minerals. This can cause fatigue, dizziness, mental disabilities, physical disabilities and even death. Rapid weight loss can also be accompanied by dehydration. When diets causing rapid weight loss restrict carbohydrates and calories, it can lead to fluid loss, according to the National Institutes of Health. If you suffer enough fluid loss, your body eventually experiences dehydration, which can lead to cramping, cerebral edema, seizures, kidney failure and even death.[28] When Lee visited Dr. Au in Hong Kong in 1972 to complain of his profuse sweating, acne, weight loss, sunken cheeks and anxiety, Dr. Au, further to the surgery that Lee had to remove the axillary sweat glands, advised Lee to eat a balanced and proper diet. It is not clear if Lee followed this advice. He probably didn't.

Did Bruce Lee die from an epileptic seizure?

In 2006 an article appeared in The Guardian Newspaper which reported that a doctor had put forward the proposition that Bruce Lee had died of an condition referred to as 'Sudden Unexpected Death in Epilepsy' or SUDEP. According to the report, '….the myth of Bruce Lee's demise in Hong Kong in 1973 may finally have been solved. "The death of Bruce Lee, coming at such a young age and in the peak of physical fitness, has given rise to much speculation," said James Filkins, at Cook County medical examiner's office in Chicago. "Almost as soon as Lee died rumors began to surface." The official cause of Lee's death was recorded in the autopsy report as cerebral edema, or brain swelling. This was supposedly due to his hypersensitivity to a painkiller called Equagesic that he had taken that day. But further research suggests the kung fu idol may have died from an epileptic condition first recognized more than 20 years after his death. The theories began on July 20 - the day he died - when he had been planning to meet his producer, Raymond Chow, and the Australian actor George

Lazenby, of James Bond fame, to try to persuade Lazenby to appear in his new film, Game of Death. He never made it to dinner. The official explanation has never satisfied Lee's fans. "Like James Dean, a death of someone so young gave cult status. He was known as the fittest man alive," said Brian Harrison, head of the Bruce Lee fan club, brucelee.org.uk. Some speculated that he had been murdered by Hong Kong triads for refusing to pay protection money or by US gangsters for refusing to work in Hollywood. A minor actress, Betty Ting Pei, was rumored to be Lee's mistress, something she has always denied. But some suggest he died while making love to her. None of these explanations washes with Dr Filkins. The autopsy report recorded no evidence of physical injury or street drugs in Lee's system apart from marijuana. There were low levels of the painkiller. Dr Filkins thinks the official explanation is also wrong. Drug reactions tend to involve an anaphylactic reaction in which the victim's neck swells, he told the annual meeting of the American Academy of Sciences in Seattle. Instead, he thinks Lee died of a condition called sudden unexpected death in epilepsy (SUDEP), which was only recognized in 1995. The condition involves a seizure which stops the heart or lungs. It kills around 500 people a year in the UK, is most common in men aged 20-40 and can be brought on by lack of sleep and stress. "Lee was under a great deal of physical and mental distress at the time," said Dr Filkins. SUDEP, or sudden unexpected death in epilepsy, accounts for 5% to 30% of deaths in patients with epilepsy. Incidence in the general epileptic population ranges from one in 370 to one in 1,100. The cause of death is thought to be a seizure-induced irregular heartbeat or respiratory arrest. The condition kills about 500 people a year in the UK, and is most common in men aged between 20 and 40.[29]

Personally, I am puzzled by the assertion that he died from SUDEP. The article states, 'Drug reactions tend to involve an anaphylactic reaction in which the victim's neck swells, he told the annual meeting of the American Academy of Sciences in Seattle.' Well, reportedly Lee's face was swollen '....like a watermelon' at the time of his death.[30] Which then would support a hypersensitivity reaction as being the cause of his death. I have further objections to this hypothesis concerning SUDEP and my own view is that we should rule it out as the primary cause of his death. Each one of us has a specific 'seizure threshold' and this can be altered in various ways, making us more vulnerable to seizures. The World Health Organization states, 'Epilepsy is a *chronic* disorder of the brain that affects people in every country of the world. It is characterized by *recurrent* seizures - which are physical reactions to sudden, usually brief, excessive electrical discharges in a group of brain cells. Different parts of the brain can be the site of such discharges.'(Author's italics) 'Seizures may result from many disorders affecting the brain. It can be a disease characterized *solely by seizures (i.e. epilepsy)*, or the seizures may be *another sign of other diseases*, e.g. metabolic disturbances (hypoglycemia, too low blood sugar), sodium and water abnormalities, calcium abnormalities etc. Seizures may also result from diseases of the brain, such as stroke, tumor, lupus, infection (meningitis) and others.'(Author's italics) Again, unlike some of those who make pronouncements about these matters, I only do so on the basis of sourced and referenced material on which I build my arguments. Let us begin by defining what SUDEP actually is. To do this I accessed the website of the SUDEP Aware, which is the organization that is tasked with raising awareness of this condition. It states on there, very clearly, 'SUDEP refers to the unexplained death of an individual, *with a diagnosis of epilepsy*, who dies suddenly, in benign circumstances, *without a structural* or toxicological cause for death being found at autopsy.'[31] (Author's italics) We also have another definition on the website, 'Sudden, unexpected, witnessed or unwitnessed, nontraumatic and nondrowning death in patients with epilepsy, with or without evidence for a seizure and excluding documented status epilepticus, in which postmortem examination does not reveal a toxicologic or *anatomic* cause for death.' (Author's italics) The anatomic cause for death in Bruce Lee's case was a swollen or edematous brain. Bruce Lee did suffer prior to his death at least one epileptic seizure, that is on May 10. That is why he was prescribed Dilantin. The website of

SUDEP Aware also tells us, 'The precise mechanism, or cause, of death is, as yet, not understood. Most sudden deaths of people with epilepsy are unwitnessed and this makes it difficult to determine what, exactly, occurs in the last moments of life. By definition, *the post mortem does not reveal a cause of death suggesting that the terminal event is due to disturbance of function, not structure.'* Again, this is evidence *against* a diagnosis of SUDEP because at Bruce Lee's death there was a disturbance of structure and not of function. That is, his brain had swelled, increased intracranial pressure and the shifting and compression of the brain structure would then have led to his rapid death. In Bruce Lee's case what had killed him was completely obvious and objectively apparent at autopsy. The only question is, what had caused the cerebral edema because that in itself is symptomatic of an underlying condition? And if he had died of an epileptic seizure, did Betty Ting report this or see this? She was with him in the apartment. I just find it difficult to believe that she would not have witnessed or have reported that Lee had had a seizure if one had taken place.

Of interest is the following post to a website that deals with epilepsy:

'My brother recently passed away from what appears to have been a bad seizure that cut off oxygen to his brain. He was 36 yrs old. That morning he got 2 seizures. He "came to" from the 1st one but not from the second. He was transported to a local hospital & given a CT scan that revealed 75% brain damage. He was then transported to a bigger hospital 45 minutes away & given another CT scan. He spent a week there & the main focus was the cerebral edema he sustained. The doctor at the 2nd hospital told us they compared the 2 CT scans & that my brother's brain swelled very rapidly from one hospital to the next. At the 2nd hospital his brain was already swollen to capacity. He said it was something he'd never seen before in all his years of medical practice & one he could not explain. As he put it, "It's a big mystery." Our family refused an autopsy even though it may have given clues as to what actually happened so now we may never know if something other than an epileptic attack caused his death. My questions to all of

you: have you ever heard of very rapid cerebral edema from an epileptic attack? [32]

The above is interesting and I'll repeat the following sentences, *'The doctor at the 2nd hospital told us they compared the 2 CT scans & that my brother's brain swelled very rapidly from one hospital to the next. At the 2nd hospital his brain was already swollen to capacity. He said it was something he'd never seen before in all his years of medical practice & one he could not explain. As he put it, "It's a big mystery."* Such things occur in medicine, and cannot always be explained. However the problem is, as in the case of Bruce Lee, with these gaps in our knowledge attempts are sometimes made to fill them by such claims as poisoning. Individuals often assume that *everything* in medicine is predictable and explicable. It is not. And conspiracy theorists often jump in to attempt to fill in the gaps with claims such as poisoning or murder. Sometimes we simply do not know why the body responds or behaves in the way that it does, but to admit that we don't know is better than to attempt to explain everything with hypotheses that are patently not verifiable. And we must look at the probabilities of all events and consider the most probable cause.

So if Lee had a face that was swollen 'like a watermelon' and if we accept this as being angioedema that occurred in conjunction with the cerebral edema, then this weighs towards a hypersensitivity reaction rather than death due to SUDEP. In any case signs indicating death due to anaphylaxis are not always apparent post mortem. One study in the medical literature states:

'The 56 deaths studied included 19 reactions to bee or wasp venom, 16 to foods, and 21 to drugs or contrast media. Death occurred within one hour of anaphylaxis in 39 cases. Macroscopic findings included signs of asthma (mucous plugging and/or hyperinflated lungs) (15 of 56), petechial hemorrhages (10 of 56), pharyngeal/laryngeal edema (23 of 56), *but for 23 of 56 there was nothing indicative of an allergic death*. Mast cell tryptase was raised in 14 of 16 cases tested; three of three tested had detectable IgE specific for the suspected allergen.' [33] So according to the newspaper article above, 'Drug reactions tend to involve an anaphylactic reaction in which the victim's neck swells....' Neck swelling is only one possible sign of a drug hypersensitivity reaction. One study states that, '....patients randomized to Omapatrilat experienced a 3-fold higher rate of angioedema than those taking Enalapril [prescribed medications] (2.17% vs. 0.68%). Angioedema with Omapatrilat was most often characterized by swelling of the lip (53%), face (32%), tongue (28%), neck (21%) and eyelids (16%), a pattern similar to that of Enalapril.[34]

There is another interesting thing to consider. Bruce Lee used marijuana, as we know. In the 19th century, marijuana was actually used to treat epilepsy.[35] However, little medical attention was subsequently given to its possible antiepileptic effects. Little is known about the extended effects of marijuana or its constituents on the brain. Short-term use of marijuana can decrease alpha amplitude [a type of brain wave] and frequency, sleep duration, and rapid eye movement (REM) sleep. However, within an average of 10 days of continued administration, these functions returned to normal.[36] Further, there is a dose-dependent effect of cannabinoids on CNS (central nervous system) excitability, with low doses producing activation and high doses reducing electrical activity. The main problem of cannabinoids used as antiepileptic drugs concerns the separation of their psychoactive from anticonvulsant activity. However, the cannabinoid metabolite: CBD, devoid of psychotropic actions, has been reported to possess anticonvulsants effects *comparable with those of Phenytoin* (based on similar spectra of anticonvulsant activity) in mice with electroshock convulsions. Moreover, clinical trials with CBD were undertaken in patients with complex partial seizures with secondary generalization. The data revealed that CBD in oral doses of 200–300 mg/day is, in fact, effective against this form of epilepsy. One epidemiologic study of illicit drug use and new-onset seizures found that marijuana use appeared to be a protective factor against first seizures in men.[37] (Author's italics.) So, and here is a further interesting point, Phenytoin is an anticonvulsant as the article states, and it is known also by the name of Dilantin, the medication that Lee was prescribed after his medical review due to the May 10 collapse and seizure. So again, the supposedly toxic effects of marijuana don't necessarily hold up. I am not advocating illicit substance use at all, however here we have clear evidence that the use marijuana is a protective factor against seizure and therefore on the night of his death his seizure threshold, due to this drug in his system, would have been increased, making him less prone to seizures. Even though he had not taken his prescribed anticonvulsant medication.

Chapter 6
Killed by a little pain-killing tablet?

Each tablet of **Equagesic**, *for oral administration, contains 200 mg meprobamate and 325 mg aspirin. Chemically, meprobamate is 2-methyl-2-propyl-1,3- propanediol dicarbamate. Its molecular formula is C9H18N2O4 with a molecular weight of 218.25.Chemically, aspirin is benzoic acid 2-(acetyloxy)-. Its molecular formula is C9H8O4 with a molecular weight of 180.16. It occurs as an odorless white, needle like crystalline or powdery substance. Meprobamate is a carbamate derivative which has been shown (in animal and/or human studies) to have effects at multiple sites in the central nervous system, including the thalamus and limbic system. Aspirin is a nonnarcotic analgesic with antipyretic and anti-inflammatory properties. It is indicated for use as an adjunct in the short-term treatment of pain accompanied by tension and/or anxiety in patients with musculoskeletal disease. Clinical trials have demonstrated that in these situations relief of pain is somewhat greater than with aspirin alone. Equagesic is not intended for use longer than 10 days. It's allergic or idiosyncratic reactions include: Severe hypersensitivity reactions, including anaphylaxis, angioneurotic edema, anuria, asthma, bronchospasm, bullous dermatitis, chills, erythema multiforme, exfoliative erythroderma, laryngeal edema, oliguria, proctitis, purpura, Stevens-Johnson syndrome, stomatitis, and urticaria. Milder reactions are characterized by an itchy, erythematous maculopapular, or urticarial rash which may be generalized or confined to the groin. Other reactions have included acute* *nonthrombocytopenic purpura, adenopathy, cross-sensitivity between meprobamate/ mebutamate and meprobamate/carbromal, ecchymoses, eosinophilia, fixed-drug eruption with cross-reaction to carisoprodol, leukopenia, peripheral edema, and petechiae. And it's central nervous system potential adverse effects include: Agitation, ataxia, **cerebral edema**, coma, confusion, dizziness, drowsiness, dysphoria, euphoria, fast EEG activity, headache, impairment of visual accommodation, lethargy, overstimulation, paradoxical excitement, paresthesias, sedation, slurred speech, subdural or intracranial hemorrhage, seizures.[1]*

'On July 20, 1973, Bruce had a minor headache. He was offered a prescription painkiller called Equagesic. After taking the pill, he went to lie down and lapsed into a coma. He was unable to be revived. Extensive forensic pathology was done to determine the cause of his death, which was not immediately apparent. A nine-day coroner's inquest was held with testimony given by renowned pathologists flown in from around the world. The determination was that Bruce had a hypersensitive reaction to an ingredient in the pain medication that caused a swelling of the fluid on the brain, resulting in a coma and death.'[2] As we have detailed above in the chapter on cerebral edema, in Lee's case it was the brain itself that became swollen with fluid.

However, on to the issue of the cause of his death, the cerebral edema. The medical literature tells us that, 'Adverse drug reactions are not at all unusual and many fatalities are recorded every year. Hypersensitivity reactions represent about one third of all adverse drug reactions. Adverse drug reactions affect10–20% of hospitalized patients and more than 7% of the general population. Severe reactions including anaphylaxis, drug hypersensitivity syndromes, Stevens Johnson syndrome and toxic epidermal necrolysis are also associated with significant morbidity and mortality. Although several risk factors have been identified, their clinical importance has not been fully understood. Future progress in immunogenetics and pharmacogenetics may help identify populations at risk for specific types of reactions.' Further, 'The most commonly implicated drugs were diuretics, non-steroidal anti-inflammatory drugs (NSAIDs) and sedatives. In Australia, 2–4% of all hospital admissions were considered to be medication-related and this figure substantially increases with the patients' age. Again, NSAIDs were the second on the list of implicated drugs after anticoagulants.' And what is one of the medications in the NSAID group? Surprise, surprise, it's Aspirin. Which Bruce Lee took on the night of his death in the Equagesic compound medication.' Anaphylactic shock is one of the severe reactions commonly associated with drug allergy fatalities. It is usually an IgE-mediated reaction and is the most frightening and potentially lethal allergic event. However, non-IgEmediated anaphylaxis, like many NSAID-induced reactions may also be equally dangerous. In the retrospective study by Kemp from 266 reported cases of anaphylaxis from a private allergy practice in Memphis, drugs (20%) were the second most recognizable cause of reactions with NSAIDs responsible for half of those.' Further, 'The dosage of the drug and the mode of administration influence the frequency of the reactions. It appears that intermittent and repeated administrations can be more sensitizing than an uninterrupted treatment. This is supported by a recent publication from Cetinkaya who studied 147 children who had received b-lactams at least three times in the preceding 12 months without allergic reaction. A 10.2% frequency of positive skin tests to penicillin was found and the author concluded that frequent use of b-lactam antibiotics leads to sensitization. Pichichero, however, found no difference in the frequency of previous b-lactam treatments in skin test negative and skin test positive children.[3]

The immune system is an integral part of human protection against disease, but the normally protective immune mechanisms can sometimes cause detrimental reactions in the host. Such reactions are known as hypersensitivity reactions, and the study of these is termed immunopathology. The traditional classification for hypersensitivity reactions is that of Gell and Coombs and is currently the most commonly known classification system. As we have discussed above, the system is not universally accepted as the final classification structure, and many researchers feel it requires revision; however we will use it for now. It divides the hypersensitivity reactions into the following 4 types:

>**Type I reactions** (i.e. immediate hypersensitivity reactions) involve immunoglobulin E (IgE)–mediated release of histamine and other mediators from mast cells and basophils.
>**Type II reactions** (i.e. cytotoxic hypersensitivity reactions) involve immunoglobulin G or immunoglobulin M antibodies bound to cell surface antigens, with subsequent complement fixation.
>**Type III reactions** (i.e. immune-complex reactions) involve circulating antigen-antibody immune complexes that deposit in postcapillary venules, with subsequent complement fixation.
>**Type IV reactions** (i.e. delayed hypersensitivity reactions, cell-mediated immunity) are mediated by T cells rather than by antibodies.

As we have stated some medical researchers believe this classification system may be too general and favor a more recent classification system proposed by Sell et al. This system divides immunopathologic responses into the following 7 categories:

>Inactivation/activation antibody reactions
>Cytotoxic or cytolytic antibody reactions
>Immune-complex reactions
>Allergic reactions
>T-cell cytotoxic reactions
>Delayed hypersensitivity reactions
>Granulomatous reactions

This system accounts for the fact that multiple components of the immune system can be involved in various types of hypersensitivity reactions. For example, T cells play an important role in the pathophysiology of allergic reactions. In addition, the term immediate hypersensitivity is somewhat of a misnomer because it does not account for the late-phase reaction or for the chronic allergic inflammation that often occurs with these types of reactions.

Allergic reactions manifest clinically as anaphylaxis, allergic asthma, urticaria, angioedema, allergic rhinitis, some types of drug reactions, and atopic dermatitis. These reactions tend to be mediated by IgE, which differentiates them from pseudoallergic (formerly called anaphylactoid) reactions that involve IgE-independent mast cell and basophil degranulation. Such reactions can be caused by iodinated radiocontrast dye, opiates, or vancomycin and appear similar clinically by resulting in urticaria or anaphylaxis. Patients prone to IgE-mediated allergic reactions are said to be atopic. Atopy is the genetic predisposition to make IgE antibodies in response to allergen exposure.

Currently allergy is considered to be synonymous with hypersensitivity in meaning. They usually refer to type 1 immediate hypersensitivity, mediated by specific IgE antibodies in genetically predisposed individuals and resulting in symptoms characteristic of eczema, urticaria, rhinitis, asthma and anaphylaxis, although it is noted that several types of allergic states encompass all the mechanisms described by Gell and Coombs.[4]

To present a further overview of hypersensitivity (or allergy) reactions we can say they literally define an exaggerated response of the immune system to antigen challenge, that is harmful to the organism itself. Although the basic phenomenology of most types of hypersensitivity reactions was established at the end of the 19th century and in the early years of the 20th century (Koch's phenomenon, Richet and Portier's anaphylaxis, Arthus' phenomenon and serum sickness), it was only in 1963 that Patrick Gell and Robin Coombs produced a comprehensive classification of hypersensitivity reactions according to their underlying immune effector mechanism. There is a significant difference between Type I hypersensitivity and the other types of hypersensitivity caused by antibodies (Type II and III), namely the fact that Type I reactions occur only in a proportion of the subjects exposed to the agent in question (the so-called atopic individuals). The majority of type II and III reactions, instead, occur in all individuals. For example haemolytic transfusion reactions (HTR) occur whenever a blood transfusion between ABO incompatible individuals is carried out, the haemolytic disease of the newborn occurs whenever a mother Rh- produce an antibody response to Rh+, fetal RBC and serum sickness occurs whenever repeated and substantial doses of foreign serum is injected in patients for therapeutic purposes, as it occurred in the early days of serotherapy. The common feature of Type I, II and III is that in all cases the antibody reactions induces cell or tissue damage (hence it appears justified that Gell and Coombs define all these reactions Hypersensitivity) but in the majority of Type II and III reactions there is no individual susceptibility and/or exaggerated response and, in the case of the HTR due to ABO incompatibility, there is not even a first sensitization phase.[5]

Enter the Professor – The British Pathologist appears.

In order to obtain what might be considered a definitive opinion Professor Donald Teare was summoned from the United Kingdom to appear at the inquest on the death of Bruce Lee. The September 20, 1973 edition of the New Straits Times stated that Professor Teare ruled out cannabis (marijuana) as the cause of death of the Kung Fu film star. The article went on, 'The testimony came from Professor Robert Donald Teare, Professor of Forensic Medicine at the University of London, who said that in the past 35 years, he had performed more than 90,000 post-mortems and was dealing with more than 250 drugs fatalities every year. Prof. Teare said Lee died of an acute cerebral edema (congestion and swelling of the brain) and that he believed that he edema was caused by "hypersensitivity to meprobamate or to aspirin or to a combination of the two." Both meprobamate and aspirin are components of a pain-killing medication that, according to earlier testimony, Lee took a short time before he collapsed and died.'

Tom Bleecker in his book correctly queries the reported number of autopsies that the Professor was reported to have completed. The newspaper reporter wrote above states that Professor Teare had performed more that 90,000 post mortems. That is an extraordinary amount and would have required the professor to have completed on average 7 autopsies each and every day of his 35 year career. Wikipedia is much more modest in claiming that he actually 'had overseen over 1,000 autopsies.'[6] It also states in Wikipedia in an article on the Professor that, 'In 1973, Teare carried out the autopsy on Bruce Lee.' It gives the source of this as the September 20, 1973 New Straits Times article quoted above however the article does not directly state that Professor Teare carried out the autopsy. So just like Bruce Lee there is a lot of misinformation surrounding this great man of medicine, but there is no doubt that he was a first rate pathologist and clinician with vast experience.

Professor Teare was a forensic heavyweight with an extraordinary profile and reputation. He was born 1 July 1911, and educated at King William's College on the Isle of Man, and Gonville and Caius College, Cambridge. He trained at St George's Hospital, London from where he qualified in 1936. Teare began his career as a lecturer in forensic medicine at St Bartholomew's Hospital Medical College. In 1963 he became reader and eventually professor of forensic medicine at Charing Cross Hospital Medical School, a post he held until retirement in 1975. Teare was also a lecturer at the Metropolitan Police College, Hendon, and served as President of the Medical Defence Union. He was also a Fellow of the Royal College of Physicians and of the Royal College of Pathologists. Teare also served as President of the British Association of Forensic Medicine.[7]

So, Equagesic. The reported culprit in Lee's death. The adverse reaction profile of Equagesic is not pretty. It has an extensive number of potential side effects. The information that I sourced the side effects profile from, as per the usual style, divided these up relating to specific bodily systems. Relating to the central nervous system, which is of interest to us here we have the following: Agitation, ataxia, *cerebral edema*, coma, confusion, dizziness, drowsiness, dysphoria, euphoria, fast EEG activity, headache, impairment of visual accommodation, lethargy, overstimulation, paradoxical excitement, paresthesias, sedation, slurred speech, subdural or intracranial hemorrhage, seizures, vertigo, and weakness.[8] (Author's italics.) In relation to 'Allergic or idiosyncratic reactions' in its side effects profile we have: Severe hypersensitivity reactions, including anaphylaxis, angioneurotic edema, anuria, asthma, bronchospasm, bullous dermatitis, chills, erythema multiforme, exfoliative erythroderma, laryngeal edema, oliguria, proctitis, purpura, Stevens-Johnson syndrome, stomatitis, and urticaria. Milder reactions are characterized by an itchy, erythematous maculopapular, or urticarial rash which may be generalized or confined to the groin. Other reactions have included acute nonthrombocytopenic purpura, adenopathy, cross-sensitivity between meprobamate/ mebutamate and meprobamate/carbromal, ecchymoses, eosinophilia, fixed-drug eruption with cross-reaction to carisoprodol, leukopenia, peripheral edema, and petechiae.' An idiosyncratic reaction is an unusual *individual* reaction to a food or drug. Or otherwise stated, idiosyncratic drug reactions may be defined as adverse effects that cannot be explained by the known mechanisms of action of the offending agent, do not occur at any dose in most patients, and develop mostly unpredictably in susceptible individuals only.[9]

Aspirin. Used every day, and usually safely, however this drug of itself can cause significant side effects and hypersentivity reactions. The mechanism of aspirin induced hypersensitivity may be related to an up regulation of the 5-lipoxygenase pathway of arachidonic acid metabolism with a resulting increase in the products of 5-lipoxygenase (such as leukotrienes). Hypersensitivity side effects include bronchospasm, rhinitis, conjunctivitis, urticaria, angioedema, and anaphylaxis with the use of aspirin. Approximately 10% to 30% of asthmatics are aspirin-sensitive. Respiratory side effects including hyperpnea, pulmonary edema, and tachypnea have occurred in patients receiving aspirin.[10] Also, in truly severe reactions to the drug aspirin even cerebral edema can occur.... causing anything from migraines to death.[11] So there is really no question and no doubt about this, cerebral edema can be caused by aspirin, and it can occur in individuals who have taken the drug without incident on previous occasions.

THE DEATH OF BRUCE LEE: A CLINICAL INVESTIGATION

At the time of Lee's death there were those who raised suspicions due to the manner in which matters unfolded. 'Reasons for suspicion: Lee was not taken to the hospital closest to his co-star's apartment; Raymond Chow announced to the press that Lee had died at his own home; Lee had acted erratically and had publicly attacked director Lo Wei on the eve of his death; traces of cannabis were found in his system; and, at the time, he was regarded as 'the fittest man in the world.' Since there was no discernible reason for his death, an official coroner's inquest was convened on September 3, 1973. The findings were as follows: that Dr. Chu, who had been called to the scene, determined that Lee was already dead, so the choice of hospital was immaterial; that Chow had refrained from mentioning Betty Ting-pei's apartment in the death announcement as a face-saving gesture to protect Lee's wife, Linda, since Lee had been romantically linked with his co-star; that Lee's attack on director Lo Wei had been simply the climax of their long-simmering hostility.'[12] This is all precisely correct. We have already stated that Bruce Lee had a temper, that's for sure. And Mr. Bleecker in Unsettled Matters insinuates that Lee was effectively out of control and going on 'rampages.' Lo Wei apparently survived the verbal altercation with Bruce, and as far as I am aware Lee was never charged with assault, or murdering anyone. And we know that many other martial artists of Lee's era had hot tempers, one of whom, '....was notorious for a temper so explosive that on occasion Ed Parker had to physically restrain him from tearing his competitor limb from limb.'[13] That is not Bruce Lee that we are referring to. I do not in any seek to condone such behavior but these men of the martial arts set of the time, including Bruce Lee, were not passive, gentle souls with an interest in flower arranging. They were warriors and some had hot tempers. So what? So have many of us. Bruce Lee was not alone in regards to this. But it was simply not indicative of itself of any medical condition and is not of itself necessarily indicative of any medication use or abuse, such as steroid use. But please correct me if I am wrong. I am not the keeper of the truth, I am only seeking it.

When Bruce Lee had the episode of cerebral edema on May 10,

1973, those responsible for his clinical care should have given clear instructions, and perhaps they did, to those in close contact with Lee on his discharge from hospital at that time. This would be to the effect that there was a possibility that this episode could repeat itself and that if he experienced a dangerous constellation of symptoms such as headache, nausea, dizziness or such symptoms he should not take or be given any medications, which might then mask the symptoms, and get himself or others were to take him to hospital for medical examination by a doctor *immediately*. On complaining of a headache he should *not* have been given any medication, especially medication that was not prescribed for him, and an ambulance should be called or Bruce Lee taken immediately to a hospital emergency room. Of course Betty Ting did not know the consequences of her giving Bruce the medication, and she has I would imagine have deeply regretted doing this on many an occasion. Like every unintended disaster we have to draw lessons from it. And the lesson here is that one should never give medications prescribed to you to anyone else. It can prove fatal. The scenario that played itself out on the night of Lee's death, as is detailed correctly and fully in 'Unsettled Matters', can only be described as disgraceful. When it was realized that he was comatose instead of a doctor or an ambulance being called the producer friend of Lee was called. He attempted to wake Bruce Lee, who was clearly non-responsive. Lee received quite a slapping from those present in a senseless attempt to wake him instead of calling an ambulance straight away.

There is no doubt that the circumstances that Bruce Lee found himself in at the time in Hong Kong can only be described as chaotic and in some ways, from our current perspective, bizarre. He had reportedly given up training altogether. For Bruce Lee training *was* his life, it was his essence. He was physically and spiritually a warrior and a martial artist. This was his life script, not in the movie sense but in the sense of being his fate and his *raison d'etre*. He was, in all probability, involved in at least one extra-marital relationship. His financial affairs were a total mess, and there was really no one looking after his financial interests. As Unsettled

Matters reveals in detail, his widow was left in a precarious financial situation, with two children to support. He was on the cusp of true and sustained international stardom and had left it to others to manage his financial affairs. Lee was not eating properly, reportedly at times resorting to carrot juice or to orange juice diets. Davis Miller has a very interesting point when he states that Lee may have suffered from an eating disorder. He had been close to death some weeks before, with the first episode of cerebral edema. And due to all of these factors, his poor diet and the stress he was under his immune system may have been compromised. According to some authorities, a compromised immune system is at increased risk of developing severe hypersensitivity or allergy reactions.[14] He probably died needlessly. His heart, his liver, his kidneys, his cerebral blood vessels were all healthy and intact as far as we can determine. He was only 32 years of age. But the sudden adverse effects of prescription medications are very, very real. And they cause thousands of deaths each year. Bruce Lee was not the first to die of such an adverse reaction to a medication, and he will not be the last. And so it was the story of a tragedy within triumph. Just as he reached his goal it was all taken away from him, and he was taken away from us. Of all things by a little tablet. Despite all that has been written about Bruce Lee, both good and bad, both true and false, he remains a hero and I remain a fan. The dynamic energy that was let loose by his life and by his achievements remains. The ethos though, as Tom Bleecker so rightly states, has been manufactured, polished, created, distorted and marketed.

But this is the irony, the same irony that was so present in the life and death of Bruce Lee. That such an extraordinarily strong and dynamic man was killed at the age of 32 by a simple medication. There are some of us who are interested in the truth about Lee, the real truth. The *precise* biography of this man. I don't have any of the resources or the fame of the Bruce Lee Foundation or Bruce Lee Enterprises, or of Mr. Bleecker, or of a publisher. And the writings of a retired nurse and independent author, who happens to live in the Philippines, will be considered perhaps by some as inconsequential. But that is not the point, because in writing this book I have learnt so much, and if others perhaps, not so much learn something, but are encouraged to think and to research and to read and to debate then that is good. And if they see my book and see that there is an alternative view, an alternative perspective, and an alternative way of framing and interpreting the facts surrounding Lee, then that is good. And it may contain, in relation to the matters we have discussed, elements of the truth that we can incorporate into a precisely detailed biography of this man including his medical history and his death. This extraordinary man, who was such a mixture of incredible fortitude and also of vulnerability. It is a truly human story, of an amazing human being, and I want the truth about him. Nothing more and nothing less.

Lee was human and like everyone had faults and frailties. But this is the man and this is the truth of being human. One of the most experienced and knowledgeable clinicians in forensic pathology in the world at that time was directly involved in the inquest in regards to the death of Bruce Lee. Professor Teare stated that Bruce Lee died of a hypersensitivity reaction to either Aspirin or Meprobamate or a combination of the two. It didn't matter that he may have had these substances on previous occasions without incident, assuming that the May 10 episode of cerebral edema was not an initial hypersensitivity reaction to the Meprobamate he had previously been prescribed. The literature tells us that. It tells us that 'About 1% of people—and 10% of those with asthma—develop a *sudden sensitivity* to aspirin, ibuprofen and other nonsteroidal anti-inflammatory drugs (NSAIDs).[15] Experts say these reactions—which may include swelling of the lips, tongue, hands and feet, as well as hives and other rashes—may happen because the NSAID doesn't fully block the body's inflammation-causing chemicals, and some overproduce instead. Researchers aren't sure why this suddenly occurs in people who have taken NSAIDs without problems for years, although stress and genetics may play roles.'[16] (Author's italics) Or this, 'Allergic reactions to foods and medicines can develop suddenly without warning, even if the particular substance has been consumed before without any difficulty. Aspirin allergy CAN (Capitalization as it appears in the sourced article –

author) begin unexpectedly but it is also important to rule out another kind of adverse drug reaction such as a predictable side effect or an idiosyncratic response; these reactions can occur inadvertently and not every time.'[17] Or if you still don't believe me, this, 'Allergic problems may arise the first time you take aspirin. Or allergy may hit suddenly after you have been taking Aspirin without problems for many years. It even can hit after age 60.'[18] And so it doesn't matter if Bruce Lee took Aspirin or Meprobamate before without incident. This in now way guarantees that a person will never suffer sometimes serious or possibly fatal reactions to the medication in the future.

And so his was a completely natural death. Such things have happened time and time again, and I believe Professor Teare and bow to his expertise and his experience in such matters. The truth of the matter is simple and mundane. The rumor mill and the conspiracy theories will continue no doubt, and that's because there are those who enjoy a good conspiracy theory and who deny the truth, and those who make money from it. I am a researcher and I am seeking the truth, and my integrity demands that I go where the facts take me. This is the way I write and this is the way I do my research. The conspiracy, murder and poisoning theorists may continue on their way. However our paths are certainly divergent. Bruce Lee died of natural causes. Of an adverse drug reaction to a prescription medication. No one out there has put up any credible, hard evidence to the contrary. There is nothing that will change that, and there is nothing that can bring him back. It is our task and our duty to accept the truth.

Rest in peace, Bruce Lee. May your soul and your spirit be free.

Chapter 7
The enduring legacy of his achievements.

Lee was named among TIME Magazine's 100 Most Important People of the Century, as one of the greatest heroes & icons, as an example of personal improvement through, in part, physical fitness, and among the most influential martial artists of the twentieth century. This is why his biography, and his medical history, requires that we examine and document everything in the most detailed and meticulous way possible. It is my own view that the Bruce Lee Foundation, or whoever it is that currently holds medical records relating to Bruce Lee, should place these in a secure facility and nominate a doctor or a nurse as a custodian of these records and allow them to be properly catalogued and examined systematically. They should be able to be accessed by serious clinical researchers interested in this aspect of his life, his health and medical history. He is far too important a figure to let all this information slip by. However the task needs to be done properly, with proper cataloging, proper referencing and proper interpretation of whatever information is there. We seek the truth, however mundane that may be, in order to construct a detailed and accurate and fully referenced biography. We owe this to Lee and to history.

After his death Lee's widow stated, "Bruce was always good to me. I could not have a complaint in the world. I could not wish for a better husband." Lee was also a loving and doting father to his two children. He liberated the Asian mindset from one of being subservient to one of being in control and of being on equal terms with the Westerner, with the white man, and with other nations that had an imperialistic past such as the United Kingdom and Japan. He was an extraordinarily intelligent man, and a revolutionary in the martial arts. And he was fallible, vulnerable, and all too human.

It is now for others to judge what I have written. I hope a rational and logical debate will continue, and in the way that I have critiqued others who hold certain views and opinions, I hope others will do this to my book too. In concluding I can only say, for those readers out there, think, research, review and make up your own mind. You be the judge as to where the truth lies.

And may the memory of Bruce Lee and his extraordinary life and work and dedication and talent live on to keep us enthralled and inspired. Enjoy your journey!

About the Author

Duncan Alexander McKenzie R.N. is a retried nurse, and an independent author and researcher. He resides in the Philippines and apart from writing enjoys reading, trail bike riding, chess, German language and German history. He is available for commissioning for specific research projects relating to any clinical matters or clinical investigation, or review of clinical information as may be documented in books or publications. He can be contacted for this purpose at:

researchnurse55@yahoo.com.ph

Other books by
Duncan Alexander McKenzie

Death and Afterlife: The Philippine Experience

The Unlucky Country. The Republic of the Philippines in the 21st Century.

The 4S Handbook. Selective Sound Sensitivity Syndrome.

Pedro Mendoza's Violin.

The Five Keys to the Kingdom of the Apocalypse.
A Techno-Horror Novel.

References and Notes.

Foreword

1) Wikipedia
2) 'Unsettled Matters The Life and Death of Bruce Lee' by Tom Bleecker
3) Ibid.
4) 'The Tao of Bruce Lee' as quoted in 'Interviews regarding Bruce Lee' with Davis Miller http://www.bruceleedivinewind.com
5) 'Unsettled Matters The Life and Death of Bruce Lee' by Tom Bleecker
6) 'The Legacy of Bruce Lee and Jeet Kune Do in the worlds of boxing and MMA.' April 08, 2011. http://www.proboxing-fans.com/the-legacy-of-bruce-lee-and-jeet-kune-do-in-the-worlds-of-boxing-and-mma_040811
7) Referencing' https://history.colorado.edu/undergraduates/paper-guidelines/referencing
8) Wikipedia
9) 'Concubines of Ancient China'

Chapter 1

1) http://www.allbrucelee.com/article/mystery_of_bruce_lee.

htm 'The Mystery of Bruce Lee's Death' by Jake Seal
2) Wikipedia
3) Ibid.
4) Ibid.
5) 'Commotio Cordis' by Barry J. Maron, M.D., and N.A. Mark Estes, III, M.D The New England Journal of Medicine
6) Ibid.
7) Ibid.
8) 'Where is the carotid sinus located?' American Association of Anatomists
9) 'Catecholamines' Medline Plus http://www.nlm.nih.gov/medlineplus/ency/article/003561.html
10) Knight 1996
11) Forensic Pathology (2nd. edition). Bernard Knight, London 1996. Chapter 'Fatal Pressure on the Neck', 361-389
12) Casper's Forensic Medicine, 1862
13) 'Pressure on the Neck'. Francis E. Camps, A.C. Hunt. Journal of Forensic Medicine 6 (1959), 116-135
14) http://www.datenschlag.org/howto/atem/english/csr.html
15) 'Traumatic Brain Injury and Increased Intracranial Pressure' by Dimitri P. Agamanolis M.D. http://neuropathology-web.org/chapter4/chapter4cHerniations.html
16) 'Touch of Death' Wikipedia
17) 'Lectures on Medical Jurisprudence' The Lancet, September, 1837
18) 'Forensic Medicine' http://www.forensicmedicine.ca/Forensics/Feigned-Diseases.html
19) http://www.bruceleedivinewind.com/tombleecker.html
20) Ibid.

Chapter 2

1) http://www.livestrong.com/article/86678-causes-brain-swelling by Dr. Franchesca Vermillion

2) 'Brain Herniation' http://missinglink.ucsf.edu/lm/ids_104_cns_injury/Herniation/Herniation.html

3) 'Cerebral edema in childhood' by Alice Ackerman MD http://www.medlink.com/medlinkcontent.asp

4) http://thejns.org/doi/abs/10.3171/foc.2007.22.5.13 'Medical management of cerebral edema' by Ahmed Raslan M.D. and Anish Bhardwaj M.D.

5) http://www.webmd.com/brain/brain-swelling-brain-edema-intracranial-pressure?page=2

6) Goldberg 1987

7) http://www.scribd.com/doc/78384258/51/Types-of-Cerebral-Edema

8) http://science.jrank.org/pages/5855/Reye-s-Syndrome.html

9) 'Reye's Syndrome' by Carl Kolchak http://voices.yahoo.com/reyes-syndrome-article-save-childs-life-117784.html

10) http://www.allbrucelee.com/article/mystery_of_bruce_lee.htm 'The Mystery of Bruce Lee's Death' by Jake Seal

11) Unsettled Matters The Life and Death of Bruce Lee by Tom Bleecker

12) 'Anaphylaxis' by Stephen F Kemp, MD, FACP; Chief Editor: Michael A Kaliner, MD

13) Da Broi, U; Moreschi, C (2011, January 30). 'Post-mortem diagnosis of anaphylaxis: A difficult task in forensic medicine.' Forensic Science International 204 (1-3): 1–5

14) http://www.fsijournal.org/article/S0379-0738(10)00218-5/abstract 'Post-mortem diagnosis of anaphylaxis: A difficult task in forensic medicine'

15) Fadal RG, Nalebuff DJ, Ali M. 'The importance of total

and allergen-specific IgE measurements.' In: Johnson F, Spencer JT (Eds). Allergy: Immunology and Medical Treatment. Symposia Specialists, Miami 1980 pp 15-28

16) 'Allergy and Hypersensitivity to Fluoride' by Bruce Spittle, M.D.

17) 'The Gell–Coombs classification of hypersensitivity reactions: a re-interpretation' by T.V. Rajan. Trends in Immunology.

18) 'Allergy and Hypersensitivity to Fluoride' by Bruce Spittle, M.D.

19) 'Unsettled Matters The Life and Death of Bruce Lee' by Tom Bleecker

20) 'Drug Induced Hypersensitivity and the HLA Complex' by Ana Alfirevic and Munir Pirmohamed in the journal 'Pharmaceuticals.'

21) 'Unsettled Matters The Life and Death of Bruce Lee' by Tom Bleecker

22) Ibid.

23) http://www.findadeath.com/Deceased/l/brucelee/bruce.htm

24) 'Cerebral Edema' http://medical-dictionary.thefreedictionary.com/cerebral+edema

25) Wikipedia

26) Binder DK, Horton JC, Lawton MT, McDermott MW, March, 2004, 'Idiopathic intracranial hypertension'. Neurosurgery 54 (3): 538–51

27) Acheson JF (2006). 'Idiopathic intracranial hypertension and visual function'. British Medical Bulletin, 79-80 (1): 233–44

28) 'Cerebral Edema and its management' http://medind.nic.in/maa/t03/i4/maat03i4p326.pdf)

29) http://www.thefreedictionary.com/rampage

30) 'Unsettled Matters The Life and Death of Bruce Lee' by Tom Bleecker

31) 'Death and changes after death' by Dr. D. Rao http://forensicpathologyonline.com/index.php?option=co

m_content&view=category&layout=blog&id=49&Itemid
=75

32) 'Cerebral angioedema associated with Enalapril' British
Journal of Clinical Pharmacology, 2009 August; 68(2):
271–273

33) 'Adrenal Crisis in Emergency Medicine' by Kevin M
Klauer, DO, FACEP; Chief Editor: Erik D Schraga, M
http://emedicine.medscape.com/article/765753-overview

34) 'BUN'
http://labtestsonline.org/understanding/analytes/bun/tab/te
st

35) http://www.caresfoundation.org/productcart/pc/images/E
mergencyInstructions.pdf)

36) 'What Are the Causes of Hypernatremia?'
http://www.ehow.com/about_5082002_causes-
hypernatraemia.html

37) Journal of Neurosurgery 'Experimental study of relation
of fever to cerebral edema' by Raymond A. Clasen, M.D.,
Sylvia Pandolfi, Iris Laing, and Donald Casey, Jr.)

38) 'Drug Allergies' World Allergy Organization
http://www.worldallergy.org/professional/allergic_disease
s_center/drugallergy

39) European Medications Agency EMA/42783/2012 Rev1
http://www.ema.europa.eu/docs/en_GB/document_library
/Referrals_document/meprobamate_107/WC500120737.p
df

40) 'Dehydration Complications'
http://topics.info.com/Dehydration-Complications_3577

41) Ibid.

42) Ibid.

43) 'Dehydration' Lennox H Huang, MD, FAAP; Chief
Editor: Timothy E Corden, MD,
http://emedicine.medscape.com/article/906999-
overview#a0199

44) Journal of Neurochemistry 'Effect of Chronic
Hypernatremic Dehydration and Rapid Rehydration on

Brain Carbohydrate, Energy, and Amino Acid Metabolism in Weanling Mice' by Jean Holowach Thurston, Richard E. Hauhart, Demoy W. Schulz

Chapter 3

1) Unsettled Matters The Life and Death of Bruce Lee by Tom Bleecker

2) Blos, P. 1960, 'Comments on the psychological consequences of cryptorchidism'. Psychoanalytic Study of Childhood, 15:395-429

3) Cytryn, L, Cytryn, E., & Reiger, R., 1967, 'Psychological implications of cryptorchidism'. Journal of American. Academic Child Psychiatry, 6:131-165

4) Cytryn, et al. p. 146

5) Schiffman & Straker 1979, 'The psychological role of the testicles: a case report' Journal of American Academy of Child Psychiatry, 18:521-526

6) Hitler Krieg und Untergang Feldherr und Diktator, by John Toland, 1981

7) 'Remembering the Master' by Sid Campbell and Greglon Lee

8) 'Unsettled Matters The Life and Death of Bruce Lee' by Tom Bleecker

9) 'Physiological effects on demography: a long-term experimental study of testosterone's effects on fitness' Reed WL, Clark ME, Parker PG, Raouf SA, Arguedas N, Monk DS, Snajdr E, Nolan V, Ketterson ED,May, 2006

10) Marazziti D, Canale D, August, 2004. 'Hormonal changes when falling in love'. Psychoneuroendocrinology 29 (7): 931–6

11) Booth A, Dabbs JM, 1993. 'Testosterone and Men's Marriages'. Social Forces 72 (2): 463–477

12) 'Disorders of the Testes' Cleveland Clinic http://www.clevelandclinic.org/health/health-

info/docs/2300/2375.asp

13) 'Maternal Hormone Levels and Risk of Cryptorchidism among Populations at High and Low Risk of Testicular Germ Cell Tumors' Katherine A. McGlynn et al. http://cebp.aacrjournals.org/content/14/7/1732.full

14) 'Role of Hormones, Genes, and Environment in Human Cryptorchidism' by Carlo Foresta, Daniela Zuccarello, Andrea Garolla and Alberto Ferlin, Endocrine Reviews

15) http://www.pediatricurologybook.com/undesendedtestes.html 'Cryptorchidism' by Thomas F. Kolon

16) 'Testicular descent: INSL3, testosterone, genes and the intrauterine milieu', by Katrine Bay, Katharina M. Main, Jorma Toppari & Niels E. Skakkebæk http://www.nature.com/nrurol/journal/v8/n4/abs/nrurol.2011.23.html

17) De Muinck Keizer-Schrama SM, Hazebroek FW, Drop SL, Degenhart HJ, Molenaar JC, Visser HK 1988 'Hormonal evaluation of boys born with undescended testes during their first year of life.' Journal of Clinical Endocrinology & Metabolism 66:159–164

18) Natural Pedia 'Diuretics and Sweating' http://www.naturalpedia.com/D/Diuretics-potassium.html

19) 'The Legendary Bruce Lee' by the Editors of Black Belt Magazine

20) 'Unsettled Matters The Life and Death of Bruce Lee' by Tom Bleecker

Chapter 4

1) Biography http://bruceleefoundation.com/index.cfm/pid/10585

2) 'The Legacy of Bruce Lee and Jeet Kune Do in the worlds of boxing and MMA.' April, 08 2011 http://www.proboxing-fans.com/the-legacy-of-bruce-lee-and-jeet-kune-do-in-the-worlds-of-boxing-and-

mma_040811
3)   http://www.ewmaa.com/bruceleefacts.html 'Bruce Lee
        Fact or Fiction'
4)   Wikipedia 'Dragon: The Bruce Lee Story'
5)   Wikipedia 'Good-morning'
6)   http://straighttothebar.com/articles/2004/07/bruce_lees_ba
        ck_injury/ 'Bruce Lee's Back Injury. What Really
        Happened' by Scott Andrew Bird
7)   'Chuck Norris Explains What Really Killed Bruce Lee'
        http://www.youtube.com/watch?v=u4ekAQfN8x8
8)   http://disease.disease.com/Signs/Edema/cerebral-
        edema.html 'Cerebral Edema'
9)   The interview took place in the Temple Discussion
        (cityonfire.com's now defunct Bruce Lee site) on 3/15/01
        between regular visitors of the site. All of the questions
        and answers were compiled by JT. Original chat session
        has been slightly edited for a tighter, easier read. Robert
        Lee Interview - http://www.cityonfire.com/robert-lee-
        interview
10)   'Unsettled Matters The Life and Death of Bruce Lee' by
        Tom Bleecker
11)   Ibid.
12)   'Epidural Cortisone Injections for Sciatica From
        Herniated Disc...Beneficial?'
        http://www.medicinenet.com/script/main/art.asp?articleke
        y=602
13)   'Low Back Pain FAQ's' by David R. Gutknecht
14)   'Does an X-ray Help Determine the Cause of Your Pain?'
        by Brett Sears
        http://physicaltherapy.about.com/b/2012/02/23/does-an-x-
        ray-help-determine-the-cause-of-your-pain.htm
15)   'Pinched Nerve & Sciatic Nerve Pain'
        http://www.ehow.com/about_5525026_pinched-nerve-
        sciatic-nerve-pain.html
16)   'Low Back Pain Fact Sheet' National Institute of
        Neurological Disorders and Stroke

http://www.ninds.nih.gov/disorders/backpain/detail_back
pain.htm

17) 'Unsettled Matters The Life and Death of Bruce Lee' by
Tom Bleecker

18) Email communication between the author and Mr.
Bleecker.

19) 'Chronic Back Pain' from the Handbook of Disabilities,
University of Missouri

20) 'Unsettled Matters The Life and Death of Bruce Lee' by
Tom Bleecker

21) 'How to avoid a catabolic state'
http://www.fitday.com/fitness-articles/fitness/body-
building/how-to-avoid-a-catabolic-state.html

Chapter 5

1) 'The New Epidemic Sweeping Across America' October,
26, 2011, by Dr. Mercola
http://articles.mercola.com/sites/articles/archive/2011/10/
26/prescription-drugs-number-one-cause-preventable-
death-in-us.aspx

2) http://findarticles.com/p/articles/mi_m1200/is_n11_v152/
ai_19805660

3) 'Bennion, L. & Li, T. K. (1976). 'Alcohol metabolism in
American Indian and Whites.' New England Journal of
Medicine, 294, 9-13

4) 'Very Potent Nepali Hashish killed Bruce Lee', post 2008
on http://www.rollitup.org/politics/65236-very-potent-
nepali-hashish-killed.html

5) 'Unsettled Matters The Life and Death of Bruce Lee' by
Tom Bleecker

6) 'Axillary Sweat Gland Removal' by Lisa Lynn Sowder,
MD, FACS

7) 'You sweat, but toxins likely stay' by Chris Woolston,
Los Angeles Times, January 28, 2008

8)   http://www.bruceleedivinewind.com/death.html
9)   http://www.bruceleedivinewind.com/death.html
10) 'Have You Ever Wondered What's In a Cigarette?'
     http://www.quitsmokingsupport.com/whatsinit.htm
11) Iversen, Leslie L., PhD, FRS, 'The Science of Marijuana'
     London, England: Oxford University Press, 2000, p. 178,
     citing House of Lords, Select Committee on Science and
     Technology, 'Cannabis - The Scientific and Medical
     Evidence' London, England: The Stationery Office,
     Parliament, 1998 and sourced from:
     http://cannabisconsumers.org/reports/drugwarfacts.php
12) 'The Pharmacy of Hashish' by E. Whineray, M.P.S.
13) http://www.imdb.com/name/nm0863916/bio
14) Bill Zimmerman, PhD, Executive Director of Americans
     for Medical Rights, November, 15, 2001
15) Francis L. Young, Administrative Law Judge for the US
     Drug Enforcement Administration (DEA) wrote in his
     September 6, 1988 decision in a case attempting to
     reschedule marijuana so that it can be prescribed by
     physicians
16) Stephen Sidney, MD, Associate Director for Clinical
     Research at Kaiser Permanente, wrote in a September 20,
     2003 editorial published in the British Medical Journal
17) Thomas Geller, MD, Associate Professor of Child
     Neurology at the Saint Louis University Health Sciences
     Center, et al., 'Cerebellar Infarction in Adolescent Males
     Associated with Acute Marijuana Use,' published in the
     journal Pediatrics in Apr. 2004
     http://medicalmarijuana.procon.org/view.answers.php?qu
     estionID=000231
18) http://www.thefreedictionary.com/infarct
19) http://www.bruceleedivinewind.com/death.html
20) http://www.celebheights.com/s/Bruce-Lee-139.html
21) 'What is a man's ideal weight?
     http://www.livestrong.com/article/92110-mans-ideal-
     weight

REFERENCES AND NOTES

22) 'The People's Almanac' by David Wallechinsky & Irving Wallace

23) Wikipedia

24) Davis Miller
http://www.bruceleedivinewind.com/davismiller.html

25) 'Vitamin A Intoxication as a cause of Pseudotumor Cerebri' by George Morrice, Jr., M.D.; William H. Havener, M.D.; Frederick Kapetansky, M.D JAMA. 1960; 173 (16):1802-1805. doi:10.1001/jama.1960.03020340020005

26) Medscape 'Vitamin A Toxicity Clinical Presentation' Author: Mohsen S Eledrisi, MD, FACP, FACE; Chief Editor: George T Griffing, MD

27) 'The Side Effects of Crash Dieting' http://www.livestrong.com/article/456844-the-side-effects-of-crash-dieting

28) 'Common Health Risks of Rapid Weight Loss' by Dana George http://www.livestrong.com/article/105084-common-health-risks-rapid-weight

29) 'Epilepsy could solve mystery of kung fu legend's death' by James Randerson, The Guardian, February 25, 2006

30) 'Unsettled Matters The Life and Death of Bruce Lee' by Tom Bleecker

31) http://www.sudepaware.org/about_sudep.html

32) http://www.epilepsy.com/node/975720 'Death from Cerebral Edema'

33) 'Postmortem findings after fatal anaphylactic reactions' Journal of Clinical Pathology 2000 April; 53(4): 273–276

34) The Office of Drug Safety (ODS) has reviewed the 'Risk Management Plan for Omapatrilat' (BMS-186716-01, IND 48,035, Serial Nos. 468 and 474; Bristol-Myers Squibb Pharmaceutical Research Institute

35) Gordon and Devinsky, 2001

36) Jones et al., 1981; Hughes, 1996

37) 'Do Central Nervous System Stimulants Lower Seizure Threshold?' Raj D. Sheth and Edgar A. Samaniego

https://www.neurology.wisc.edu/publications/2008/Neuro
_12.pdf

Chapter 6

1) 'Equagesic'
   http://dailymed.nlm.nih.gov/dailymed/drugInfo.cfm?id=6
   166
2) The Bruce Lee Foundation – biography section
   http://bruceleefoundation.com/index.cfm/pid/10585
3) 'Epidemiology of hypersensitivity drug reactions' by Eva
   Rebelo Gomes and Pascal Demoly
   http://imb.usal.es/formacion/docencia/alergenos/Drogas,
   %20medicamentos,%20l%E1tex,/drugs%20allergy.pdf
4) Fadal RG, Nalebuff DJ, Ali M. 'The importance of total
   and allergen-specific IgE measurements.' In: Johnson F,
   Spencer JT (Eds). Allergy: Immunology and Medical
   Treatment. Symposia Specialists, Miami 1980 pp 15-28
5) 'Hypersensitivities'
   http://nfs.unipv.it/nfs/minf/dispense/immunology/lectures/
   files/hypersensitivities.html
6) Wikipedia
7) Ibid.
8) 'Equagesic' http://www.drugs.com/pro/equagesic.html
9) 'Idiosyncratic adverse reactions to antiepileptic drugs'
   http://www.ncbi.nlm.nih.gov/pubmed/17386054
10) 'Aspirin and Codeine Side Effects'
    http://www.drugs.com/sfx/aspirin-and-codeine-side-
    effects.html
11) 'Aspirin side effects' http://www.ibnisina.org/aspirin-
    side-effects
12) 'The People's Almanac' by David Wallechinsky & Irving
    Wallace
13) 'Unsettled Matters The Life and Death of Bruce Lee' by
    Tom Bleecker

14) 'The Immune System' by William R. Kellas
http://www.awarenessmag.com/julaug9/JA9_IMMU.HT
ML

15) 'Adult Mystery: Sudden 'Allergy' The Wall Street
Journal Health Journal August 16, 2011

16) 'Can you suddenly be allergic to Aspirin?' Dr. Alex
Martinez
https://www.healthtap.com/#user_questions/44596-can-
you-be-suddenly-allergic-to-aspirin

17) 'How to Reduce Drug Side Effects' by Diana Benzaia,
MA http://www.hss.edu/conditions_14163.asp